CAKES, CUSTARD + CATEGORY THEORY

Eugenia Cheng is Senior Lecturer in Pure Mathematics at the University of Sheffield. She was educated at the University of Cambridge and has done post-doctoral work at the Universities of Cambridge, Chicago and Nice. Since 2007 her YouTube lectures and videos have been viewed over a million times. A concert pianist, she also speaks French, English and Cantonese, and her mission in life is to rid the world of maths phobia.

CAKES, CUSTARD + CATEGORY THEORY

Easy recipes for understanding complex mathematics

EUGENIA CHENG

P

PROFILE BOOKS

First published in Great Britain in 2015 by
PROFILE BOOKS LTD
3 Holford Yard
Bevin Way
London
WC1X 9HD

www.profilebooks.com

1 3 5 7 9 10 8 6 4 2

Printed and bound in Great Britain by
Clays, Bungay, Suffolk

A CIP catalogue record for this book is available from the British Library.

ISBN 978 178125 2871
eISBN 978 178283 0825

CONTENTS

To
my parents
and Martin Hyland

In memory of
Christine Pembridge

They say mathematics is a glorious garden. I know I would certainly lose my way in it without your guidance. Thank you for walking us through the most beautiful entrance pathway.

From a student's letter to the author
University of Chicago, June 2014

PROLOGUE

Here is a recipe for clotted cream.

Ingredients

Cream

Method

1. Pour the cream into a rice cooker.
2. Leave it on 'warm' with the lid slightly open, for about 8 hours.
3. Cool it in the fridge for about 8 hours.
4. Scoop the top part off: that's the clotted cream.

What on earth does this have to do with maths?

Maths myths

Maths is all about numbers.

You might think that rice cookers are for cooking rice. This is true, but this same piece of equipment can be used for other things as well: making clotted cream, cooking vegetables, steaming a chicken. Likewise, maths is about numbers, but it's about many other things as well.

Maths is all about getting the right answer.

Cooking is about ways of putting ingredients together to make delicious food. Sometimes it's more about the method than the ingredients, just as in the recipe for clotted cream, which only has one ingredient – the entire recipe is just a method. Maths is about ways of putting ideas together, to make exciting new ideas. And sometimes it's more about the method than the 'ingredients'.

Maths is all either right or wrong.

Cooking can go wrong – your custard can curdle, your soufflé can collapse, your chicken can be undercooked and give everyone food poisoning. But even if it doesn't poison you, some food tastes better than other food. And sometimes when cooking goes 'wrong' you have actually accidentally invented a delicious new recipe. Fallen chocolate soufflé is deliciously dark and squidgy. If you forget to melt the chocolate for your cookies, you get chocolate chip cookies. Maths is like this too. At school if you write $10 + 4 = 2$ you will be told that is wrong, but actually that's correct in some circumstances, such as telling the time – four hours later than 10 o'clock is indeed 2 o'clock. The world of maths is more weird and wonderful than some people want to tell you. . .

You're a mathematician? You must be really clever.

Much as I like the idea that I am very clever, this popular myth shows that people think maths is hard. The little-understood truth is that the aim of maths is to make things easier. Herein lies the problem – if you need to make things easier it gives the impression that they were hard in the first place. Maths *is* hard, but it makes hard things easier. In fact, since maths is a hard thing, maths also makes maths easier.

Many people are either afraid of maths, or baffled by it, or both. Or they were completely turned off it by their lessons at school. I understand this – I was completely turned off sport by my lessons at school, and have never really recovered. I was so bad at sport at school, my teachers were incredulous that anybody so bad at sport could exist. And yet I'm quite fit now, and have even run the New York Marathon. At least I now appreciate physical exercise, but I still have a horror of any kind of team sport.

How can you do research in maths? You can't just discover a new number.

This book is my answer to that question. It's hard to answer it quickly at a cocktail party, without sounding trite, or taking up too much of someone's time, or shocking the gathered company. Yes, one way to shock people at a polite party is to talk about maths.

It's true, you can't just discover a new number. So what can we discover that's new in maths? In order to explain what this 'new maths' could possibly be about, I need to clear up some misunderstandings about what maths is in the first place. Indeed, not only is maths not just about numbers, but the branch of maths I'm going to describe is actually not about numbers at all. It's called *category theory* and it can be thought of as the 'mathematics of mathematics'. It's about relationships, contexts, processes, principles, structures, cakes, custard.

Yes, even custard. Because mathematics is about drawing analogies, and I'm going to be drawing analogies with all sorts of things to explain how maths works. Including custard, cake, pie, pastry, doughnuts, bagels, mayonnaise, yoghurt, lasagne, sushi.

Whatever you think maths is . . . let go of it now.
This is going to be different.

part one

MATHEMATICS

1 WHAT IS MATHS?

Gluten-free chocolate brownies

Ingredients

115 g butter
125 g dark chocolate
150 g caster sugar
80 g potato flour
2 medium eggs

Method

1. Melt the butter and chocolate, stir together and allow to cool a little.
2. Whisk the eggs and the sugar together until fluffy.
3. Beat the chocolate into the egg mixture slowly.
4. Fold in the potato flour.
5. Bake in very small individual cases at 180°C for about 10 minutes, or until they're as cooked as you want them.

Maths, like recipes, has both ingredients and method. And just as a recipe would be a bit useless if it omitted the method, we can't understand what maths is unless we talk about the *way it is done*, not just the *things it studies*. Incidentally the method in the above recipe is quite important – these don't cook very well in a large tray. In maths the method is perhaps even more important than the ingredients. Maths probably isn't whatever you studied at school in lessons called 'maths'. Yet somehow I always knew that maths was more than what we did at school. So what *is* maths?

Recipe books

What if we organised recipes by equipment?

Cooking often proceeds a bit like this: you decide what you want to cook, you buy the ingredients, and then you cook it. Sometimes it might work the other way round: you go wandering around the shops, or maybe a market. You see what ingredients look good, and you feel inspired by them to conjure up your meal. Perhaps there's some particularly fresh fish, or a type of mushroom you've never seen before, and you go home and look up what to do with it afterwards.

Occasionally something completely different happens: you buy a new piece of equipment, and suddenly you want to try making all sorts of different things with that equipment. Perhaps you bought a blender, and suddenly you make soup, smoothies, ice cream. You try making mashed potatoes in it, and it goes horribly wrong (it looks like glue). Maybe you bought a slow cooker. Or a steamer. Or a rice cooker. Perhaps you learn a new technique, like separating eggs or clarifying butter, and suddenly you want to make as many things as possible involving your new technique.

So we might approach cooking in two ways, and one seems much more practical than the other. Most recipe books are divided up according to parts of the meal rather than by techniques. There's a chapter on starters, a chapter on soup, a chapter on fish, a chapter on meat, a chapter on dessert, and so on. There might be a whole chapter on an ingredient, say a chapter on chocolate recipes or vegetable recipes. Sometimes there are whole chapters on particular meals, say a chapter on Christmas lunch. But it would be quite odd to have a chapter on 'recipes that use a rubber spatula' or 'recipes that use a balloon whisk'. Having said that, kitchen gadgets often come with handy books of recipes you can make with your new equipment. A blender will come with blender recipes; likewise a slow cooker or an ice cream maker.

Something similar is true of subjects of research. Usually when you say what a subject is, you describe it according to the thing that you're studying. Maybe you study birds, or plants, or food, or cooking, or how to cut hair, or what happened in the past, or how society works. Once you've decided what you're going to study, you learn the techniques for studying it, or you invent new techniques for studying it, just like learning how to whisk egg whites or clarify butter.

In maths, however, the things we study are also determined by the techniques we use. This is similar to buying a blender and then going round seeing what you can make with it. This is more or less backwards from other subjects. Usually the techniques we use are determined by the things we're studying; usually we decide what we want for dinner, and then get out the equipment for making it. But when we're really excited about our new blender, we go round trying to make all our dinners in it for a while. (At least, I've seen people do this.)

It's a bit of a chicken-and-egg question, but I am going to argue that maths is defined by the techniques it uses to study things, and that the things it studies are determined by those techniques.

Cubism

When the style affects the choice of content

Characterising maths by the techniques it uses is similar to defining styles of art, like cubism or pointillism or impressionism, where the genre is defined by the techniques rather than the subject matter. Or ballet and opera, where the art form is defined by the methods, and the subject matter is duly restricted. Ballet is very powerful at expressing emotion, but not so good at expressing dialogue, or making demands for political change. Cubism is not that effective for depicting insects. Symphonies are good at expressing tragedy and joy, but not very good at saying 'Please pass the salt.'

In maths the technique we use is *logic*. We only want to use sheer logical reasoning. Not experiments, not physical evidence, not blind faith or hope or democracy or violence. Just logic. So what are the things we study? We study *anything that obeys the rules of logic*.

Mathematics is the study of anything that obeys the rules of logic, using the rules of logic.

I will admit immediately that this is a somewhat simplistic definition. But I hope that after reading some more you'll see why this is accurate as far as it goes, not as circular as it sounds at first, and just the sort of thing a category theorist would say.

The prime minister

Characterising something by what it does

Imagine if someone asked you 'Who's the prime minister?' and you answered 'He's the head of the government.' This would be correct but annoying, and not really answering the right question: you've characterised the prime minister without telling us who it is. Likewise, my 'definition' of mathematics has *characterised* maths rather than telling you what it is. This is a little unhelpful, or at least incomplete – but it's just the start.

Instead of describing what maths is *like*, can we say what maths *is*? What does maths actually study? It definitely studies numbers, but also other things like shapes, graphs and patterns, and then things that you can't see – logical ideas. And more than that: things we don't even know about yet. One of the reasons maths keeps growing is that once you have a technique, you can always find more things to study with it, and then you can find more techniques to use to study those things, and then you can find more things to study with the new techniques, and so on, a bit like chickens laying eggs that hatch chickens that lay eggs that hatch chickens...

Mountains

Conquering one enables you to see the higher ones

Do you know that feeling of climbing to the top of a hill, only to find that you can now see all the higher hills beyond it? Maths is like that too. The more it progresses, the more things it comes up with to study. There are, broadly, two ways this can happen.

First there's the process of 'abstraction'. We work out how to think logically about something that logic otherwise couldn't handle. For example, you previously only made rice in your rice cooker, and then you work out that you can use it to make cake, it's just a bit different from cake made the normal way in an oven. We take something that wasn't really maths before, and look at it differently to turn it into maths. This is the reason that x's and y's start appearing – we start by thinking about numbers, but then realise that the things we do with numbers can be done with other things as well. This will be the subject of the next chapter.

Secondly there's the process of 'generalisation': we work out how to build more complicated things out of the things we've already understood. This is like making a cake in your blender, and making the icing in your blender, and then piling it all up.* In maths this is how we get things like polynomials and matrices, complicated shapes, four-dimensional space, and so on, out of simpler things like numbers, triangles and our everyday world. We'll look into this in Chapter 5.

These two processes, abstraction and generalisation, will be the subject of the next few chapters, but first I want to draw your attention to something weird and wonderful about how maths does these two things.

* Mathematical generalisation isn't the same as the kind where you go round making sweeping statements about things, but we'll come to that later.

Birds

They are not the same as the study of birds

Imagine for a second that you study birds. You study their behaviour, what they eat, how they mate, how they look after their young, how they digest food, and so on. However, you will never be able to build a new bird out of simpler birds – that just isn't how birds are made. So you can't do generalisation, at least not in the way that maths does it.

Another thing you can't do is take something that isn't a bird, and miraculously turn it into a bird. That also isn't how birds are made. So you can't do abstraction either. Sometimes we realise we've made a mistake of classification – for example the brontosaurus 'became' a form of apatosaurus. However, we didn't turn the brontosaurus into an apatosaurus – we merely realised it had been one all along. We're not magicians, so we can't change something into something it isn't. But in maths we can, because maths studies ideas of things, rather than real things, so all we have to do to change the thing we're studying is to change the idea in our head. Often, this means changing the way we think about something, changing our point of view or changing how we express it.

A mathematical example is knots.

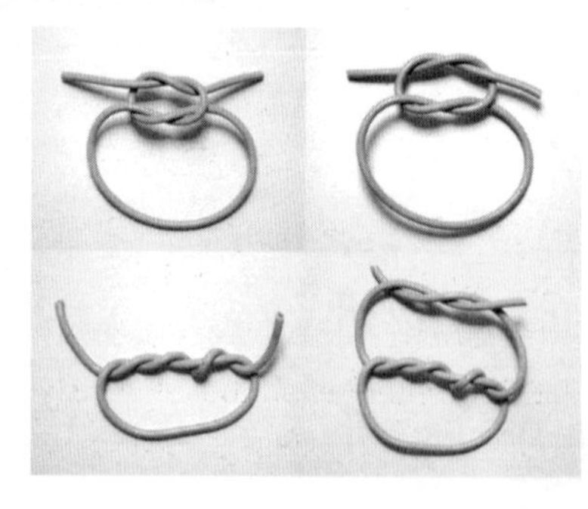

In the eighteenth and nineteenth centuries, Vandermonde, Gauss and others worked out how to think of knots mathematically, so that they could be studied using the rules of logic.

The idea is to imagine sticking together the two ends of the piece of string so that it has become a closed loop. This makes the knots impossible to make without glue, but much easier to reason with mathematically. Each one can be expressed as a circle that has been mapped to three-dimensional space. There are many techniques for studying this kind of thing in the field of *topology*, which we'll come back to later. We can then deduce things not only about real knots in string, but also about the apparently impossible ones that arise in nature in molecular structures.

Geometrical shapes are another, much older example of this process of turning something from the 'real' world into something in the 'mathematical' world.

We can think of maths as developing in the following stages:

1. It started as the study of numbers.
2. Techniques were developed to study those numbers.
3. People started realising that those techniques could be used to study other things.
4. People went round looking for other things that could be studied like this.

Actually there's a step 0, before the study of numbers: someone had to come up with the idea of numbers in the first place. We think of them as the most basic things you can study in maths, but there was a time before numbers. Perhaps the invention of numbers was the first ever process of *abstraction*.

The story I'm going to tell is about abstract mathematics. I'm going to argue that its power and beauty lie not in the answers it provides or the problems it solves, but in the *light* that it sheds. That light enables us to see clearly, and that is the first step to understanding the world around us.

2 ABSTRACTION

Mayonnaise or hollandaise sauce

Ingredients

2 egg yolks

300 ml olive oil

Seasoning

Method

1. Whisk the egg yolks and seasoning using a hand whisk or immersion blender.
2. Drip the olive oil in very slowly, while continuing to whisk.

For hollandaise sauce, use 100 g melted butter instead of the olive oil.

At some level mayonnaise and hollandaise sauce are the same – they use the same method, but with a different type of fat incorporated into the egg yolk. In both cases, the amazing near-magic properties of egg yolks create something rich and unctuous. It looks so much like magic, I never tire of watching it happen.

The similarity between mayonnaise and hollandaise sauce is the sort of thing that mathematics goes round looking for – situations where things are somehow the same apart from some small detail. This is a way of saving effort, so that you can understand how to do both things at once. Books might tell you that hollandaise sauce needs to be done differently, but I ignore them to make my life simpler. Maths is also there to make things simpler, by finding things that look the same if you ignore some small details.

Pie

• • •

Abstractions as blueprints

Cottage pie, shepherd's pie and fisherman's pie are all more or less the same – the only difference is the filling that is sitting underneath the mashed potato topping. Crumble is also very similar – you don't really need a different recipe for different types of crumble, you just need to know how to make the crumble part. Then you put the fruit of your choice in a dish, and put the crumble on top, and bake it.

Another favourite of mine is upside-down cake. You put the fruit in the bottom of the cake tin, pour the cake mix on top, and after baking it you turn it out upside down so that the fruit is on top. For extra effect you can put melted butter and brown sugar on the bottom of the cake tin first, to caramelise the fruit a bit. Of course, this works better with some fruit than others: bananas, apples, pears and plums work well. Grapes less well. Watermelon would be terrible. The same is true for crumble. Watermelon crumble? Probably not.

Savoury tarts and quiches also follow a general pattern. You bake an empy pastry case, put in some filling of your choice, and then top it up with a mixture of egg and milk or cream, before baking it again. The filling could be bacon and cheese, or fish, or vegetables – whatever you feel like.

In all these cases the 'recipe' is not a full recipe, but a blueprint. You can insert your own choice of fruit, or meat, or fillings to make your own variations, within reason.

This is also how maths works. The idea of maths is to look for similarities between things so that you only need one 'recipe' for many different situations. The key is that when you ignore some details, the situations become easier to understand,

and you can fill in the variables later. This is the process of abstraction.

As with the watermelon crumble, once you've made the abstract 'recipe' you will find that you won't be able to apply it to *everything*. But you are at least in a position to try, and sometimes surprising things turn out to work in the same recipe.

Think about the symmetry of an equilateral triangle:

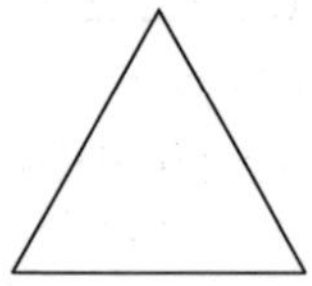

There's reflectional symmetry, and rotational symmetry. How can we describe the different symmetries without cutting out the triangle and folding it up or waving it around?

One way is that we could label the corners 1, 2 and 3,

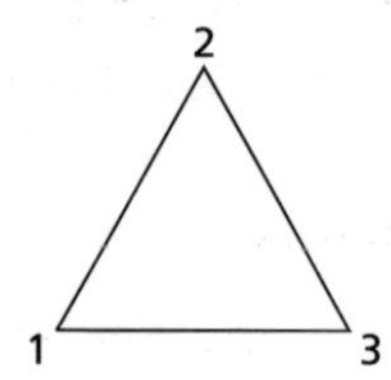

and then just talk about how the numbers get swapped around. For example, if we reflect the triangle in a vertical line, we will swap the numbers 1 and 3. Whereas if we rotate the triangle 120° clockwise we will send 1 to where 2 was, 2 to where 3 was, and 3 to where 1 was.

You can try checking that the six symmetries of the triangle correspond exactly to the six different ways of shuffling the numbers 1, 2 and 3. There are three lines of symmetry, and they correspond to swapping 1 and 3, or 1 and 2, or 2 and 3. There are three types of rotational symmetry: 120° clockwise, 240° clockwise, and the 'trivial' one where nothing moves.

This shows that the symmetry of an equilateral triangle is

abstractly the same as the permutations of the numbers 1, 2, 3, and the two situations can be studied at the same time.

Kitchen clutter

Abstraction as tidying away the things you don't need

Abstraction is like preparing to cook something, and putting away the equipment and ingredients that you don't need for the recipe, so that your kitchen is less cluttered. It is the process of putting away the ideas you don't need for the present purposes, so that your *brain* is less cluttered.

Are you better at this in your kitchen, or in your brain? (I am definitely better at it in my brain.) Abstraction is the important first step of doing mathematics. It's also a step that can make you feel uneasy because you're stepping away from reality a little bit. I never put my food processor away because it's such a hassle to move it, and I want to know that I can use it any minute now without going through the rigmarole of getting it out of the cupboard. You might feel like that about abstraction in the brain as well.

Try the following problem:

I buy two stamps for 36p each. How much does it cost?

When children do this sort of thing at primary school it sometimes get called a 'word problem', because it has been stated in words, and they're told that the first step in solving this 'word problem' is to turn it into numbers and symbols:

$$36 \times 2 = ?$$

This is a process of abstraction. We have thrown away, or ignored, the fact that the thing we were buying was *stamps*, because it didn't make any difference to the answer. It could have been apples, bananas, monkeys, . . ., the sum would still be the same, and so the answer would still be the same: 72p.

What about this one:

> *My father is three times as old as I am now but in ten years' time he will be twice as old as me. How old am I?*

Or this one:

> *I have a recipe for icing the top and sides of a 6-inch cake. How much icing do I need for the top and sides of an 8-inch cake?*

For the question about stamps you probably didn't need to write down a sum, because the answer was immediately obvious to you. However, for these last two questions, perhaps you would need to perform some abstraction to work out the answer, where you throw away the fact that you're talking about your father, or a cake and icing, and write down some sums, with numbers and symbols. We'll see what sums we get from these word problems a bit later in this chapter.

Sweets

How things that are too real don't obey mathematics

If you've ever tried teaching arithmetic to small children, you might have come up with the following problem. You try and get them to think about a real-life situation such as:

> *If grandma gives you five sweets and grandpa gives you five sweets, how many sweets will you have?*

And the child answers: 'None, because I'll eat them all!'

The trouble here is that sweets do not obey the rules of logic, so using maths to study them doesn't quite work. Can we force sweets to obey logic? We could impose an extra rule on the situation by adding '... and you're not allowed to eat the sweets'. If you're not allowed to eat them, what's the point of them being sweets? We could treat the sweets as just *things* rather than sweets. We lose some resemblance to reality, but we gain scope and with it efficiency. The point of numbers is that we can reason about 'things' without having to change the reasoning depending on what 'thing' we are thinking about. Once we know that $2 + 2 = 4$ we know that two things and another two things make four things, whether they are sweets, monkeys, houses, or anything else. That is the process of abstraction: going from sweets, monkeys, houses, or whatever, to numbers.

Numbers are so fundamental, it's difficult to imagine life without them, and difficult to imagine the process of inventing them. We don't even notice that we're making a leap of abstraction when we count things. It's much more noticeable if you watch small children struggling to do it, because they're not yet used to making that leap.

Eeny meeny miny moe

Numbers as an abstraction

I remember a wonderfully feisty mother at a primary school I was helping at. She also helped there, and remarked on how frustrating it was when other mothers competitively declared that *their* child could count up to 20 or 30. 'My son can count up to three,' she said defiantly. 'But he knows what three *is*.'

And she had a point.

When a child first 'learns to count to ten' they aren't really doing more than learning to recite a little poem, like 'Incy wincy spider climbed up the spout...' It just so happens that the little 'poem' goes:

'One, two, three, four, five, six, ... '

Then they learn that this has something to do with pointing at things, so they start pointing while reciting the 'poem', a bit haphazardly.

Next they learn that they're supposed to point at one thing per word in the poem, but they have trouble making sure they have only pointed at each thing once, so they will get rather variable answers if you ask them 'How many ducks are in this picture?' Or they might latch on to a particular number – say, six – and somehow manage to count everything as being six, no matter how many ducks there really are.

Finally they'll get the idea that they're supposed to match up the items rather precisely with the words in the poem, one item per word, no more, and no less. That is when they *really* know how to count. This is a process of abstraction, and a surprisingly profound one.

Imagine trying to do trade without knowing how to count. 'Hey, I'll trade you one sack of grain for each of your sheep,' and then you go and line up sacks of grain against sheep to make sure you really have one per sheep. Eventually you work out that it's more practical to recite a little poem while pointing at the sheep in rhythm, and do the same thing with the sacks of grain. The poem could be anything as long as you recite it exactly the same way for the sheep and for the grain. It could be 'Eeny meeny miny moe'.

Finally you make up a poem once and for all to use for all your trades, and you stick to it. And suddenly you've invented numbers. That is the process of abstraction that we don't even notice when we 'learn to count'. So we see that there is a crucial difference between simply learning the poem 'One, two, three, four, . . .' and understanding how to use it.

The baby and the bathwater

Being careful not to throw away too much

It is important, as everyone knows, not to throw the baby out with the bathwater. When we go round simplifying or idealising our situations, we must be careful not to *oversimplify* – we must not simplify our objects to the point that they've lost *all* of their useful characteristics. If we're thinking about stacking Lego blocks, for example, we can ignore what colour they are, but we shouldn't ignore what size they are, as that affects how we can stack them. But in another situation we might be using Lego bricks merely as counting blocks, in which case we can ignore their size as well.

Choosing what features to ignore should depend heavily on what context we're thinking about. This is a theme that will come back importantly later. Category theory brings context to the forefront.

Suppose you're organising an outing for 100 people, and you're hiring minibuses that can hold 15 people each. How many minibuses do you need? Basically you need to calculate

$$100 \div 15 \approx 6.7$$

but then you have to take the context into account: you can't book 0.7 of a minibus, so you'll need to round up to 7 minibuses.

* * *

Now consider a different context. You want to send a friend some chocolates in the post, and a first-class stamp is valid for up to 100 g. The chocolates weigh 15 g each, so how many chocolates can you send? You still need to start with the same calculation

$$100 \div 15 \approx 6.7$$

but this time the context gives a different answer: you can't send 0.7 of a chocolate, so you'll need to round *down* to 6 chocolates.

Heartbreak

Abstraction as simplification

After one major episode of heartbreak I was getting tired of well-meaning friends asking me for more and more details of exactly what happened, in an attempt to 'understand' it. Finally one wise friend said to me, 'It's very simple really. You've lost something you loved.' That was all anyone needed to know of the situation. She then successfully distracted me into a long discussion about how it's really more intelligent to be able to simplify things than to complicate them, even if some people think it makes you look stupid. There's a subtle difference between something that's 'simple' and something that's 'simplistic'; the latter indicates that you've missed the point, and ignored a complication that is crucial.

My friend's wisdom was a type of abstraction, abstracting heartbreak down to its very essence. Abstraction can appear to take you further and further away from reality, but really you're getting closer and closer to the heart of the matter. To get to the heart, you have to strip away clothes and skin and flesh and bone.

Road signs

Abstraction as the study of ideal versions of things

Road signs are a form of abstraction. They don't precisely depict what is going on in the road, but represent some idealised form of it, where just the essence is captured. Not every humpbacked bridge looks exactly like this:[1]

[1] Road sign images are Crown Copyright and reproduced under the Open Government Licence.

but this captures the essence of humpbacked-bridge-ness. Similarly, not all children crossing the road look exactly like this:

Nevertheless the benefits of this system are clear. It's much quicker to take in a symbol than read some words while you are driving. Also it's much easier for foreigners to understand. The disadvantage is that when you first start driving you have to learn what all these funny symbols mean. Some of them, for example

are much closer to reality than others, for example

This 'No Entry' sign is entirely abstract: it doesn't look like the thing it is representing at all. (What does 'No Entry' look like?) But it's also more important – you will probably encounter more of those in your driving life than the one warning you there might be deer crossing the road.

One side effect of the abstraction of maths is that a variety of funny symbols get used as well, for the same sorts of reasons: once you know what they mean, the symbols are much quicker to take in, and you can reserve your mathematical brain power for the more complicated parts of the maths you're supposed to be focusing on. It also makes the maths easier to understand across different languages – it's surprisingly easy to read a maths book in a language you don't know.

The most basic 'funny symbols' used in maths are the ones for normal arithmetic: $+$, $-$, $\times$, $\div$, $=$. Once you're comfortable with these symbols, it's much quicker and easier to read

$$2 + 2 = 4$$

than 'two plus two equals four'. As maths gets more and more complicated, the symbols get more and more complicated as well, with things like

$$\sum, \int, \oint, \otimes, \Leftrightarrow, \models, \ldots$$

I'm not going to explain what the more complex symbols mean here – this is just to give an idea of some of the symbols that get used. As with road signs, it makes maths look a bit incomprehensible at first, but it makes it easier in the long run.

Google Maps

The difficulty of relating the map to the reality

What is difficult about reading a map? It's not the actual reading of the map that's hard, but matching that up with reality in order to put the map to practical use. A map is an abstraction of reality. It depicts certain aspects of reality that are supposed to help you find your way around. The difficulty, in practice, is in translating between the abstraction and the reality. That is, making the link between the map and the place you're actually wandering around.

Google Maps gives us a brilliant way of moving from the abstract to the concrete, via Google Street View and GPS. Often the hardest part about using a map is working out

a. where you are in the first place, and
b. which way you're facing.

Those are the crucial pivot points between the map and the reality. GPS has sorted out the business of working out where you are and Google Street View has sorted out the business of which way you're facing, by giving us a very realistic representation of reality in the form of an actual picture of it.

Maths has to go through these steps as well. First you have to turn the reality into an abstraction. Then you do your logical reasoning in the abstract world. Then finally you have to turn that back into reality again. Different people are good at different parts of this process. But really the key part is being able to move back and forth between the abstract and the real. Still, *someone had to draw the map.*

For example, suppose you have a recipe for an 8-inch square cake, but you want to make it round instead. What size of round cake tin should you use? First you perform an abstraction to turn this 'real-life' question into a piece of maths. We want to find a circle whose area is the same as the area of the given square, which is $8^2 = 64$. Now we have to remember that the area of a circle is πr^2, where r is the radius. If we write d for the diameter of the circle (because cake tins are measured by their diameter not their radius), this means we need

$$\pi\left(\frac{d}{2}\right)^2 = 64.$$

Now we actually do the logical reasoning, manipulating the algebra to find out what the diameter d needs to be. This is the only part that's actually maths.

$$\begin{aligned}\left(\frac{d}{2}\right)^2 &= \frac{64}{\pi}\\ \frac{d}{2} &= \sqrt{\frac{64}{\pi}}\\ d &= 2 \times \sqrt{\frac{64}{\pi}}\\ &\approx \pm 9.027\end{aligned}$$

Finally we take the context into account and turn this back into reality. First of all, we don't want the negative answer because we're talking about cake tins here, so the answer needs to be a positive number. Secondly, we don't need all those decimal places – cake tins are usually only measured to the nearest inch. So the answer in reality is that we need a 9-inch round tin for our cake.

The key in maths, and with maps, is to find the most appropriate level of abstraction for the given moment. Do you need little pictures of all the buildings on a street when you're looking at a street map? Do you need to know where there is grass and where there isn't? It depends what you're using the map for, and you'll need different maps for different situations. If you're driving, then you'll want to know which streets are one-way, but that's not very relevant if you're on foot. The same is true of maths. There are different levels of abstraction available for different situations.

> What is the number 1? Here are two different ways of answering that question, at different levels of abstraction.
>
> First answer: 1 is the basic building block of counting.
>
> Second answer: 1 is the only number with the property that multiplying by it does nothing.
>
> Each of these answers is useful in different contexts. The first is for when we are most interested in adding numbers up; in mathematics this characterises numbers as something called a 'group' – a world in which we can do addition. The second is for when we are also interested in multiplying; this characterises numbers as something called a 'ring' – a world in which we can do addition *and* multiplication. The study of groups is related to the symmetry of shapes, and the study of rings is related to other aspects of the geometry of shapes. We'll come back to this later.

If you use an inappropriate map for the situation you're in, you'll get frustrated, whether it's too realistic or not realistic enough. (I dislike those street maps with pictures of buildings on in three dimensions, so that they actually obscure the lines telling you where the street goes.)

The same is true of maths – if you try and use complicated maths for a situation that doesn't call for it, you'll think the maths is pointless. It's a bit like using the Dewey decimal system on your books if you only own twenty books.

High jump

Leaps of abstraction

I was terrible at the high jump at school. Of course, I already said I was terrible at all sport, but with the high jump I failed before I started – I couldn't jump over the bar even at its lowest.

The trouble is that nobody tried to teach me what I needed to do to get myself over that bar. Other people in my class just seemed to be able to do it, as if by magic, and the rest of us were simply told to do it again. And again. And again. There are only so many times you can knock down a high-jump bar, with an audience, without feeling disillusioned and keen to give up.

Thinking about more and more abstract concepts is a bit like the high jump. You have to get yourself over a progressively higher and higher bar, and if nobody explains how to do it, you will keep knocking the bar off and want to give up. Different people reach their limit of abstraction at different levels, just as, with the high jump, people drop out at each round.

Most people are able to make the abstraction from *objects* to *numbers* and don't even notice that is a process of abstraction at all. One level where many people find they can't get over the bar any more is where the numbers turn into *x*'s and *y*'s. They can't do it, and they also can't see the point of doing it, so they get disillusioned and give up. (I never saw the point of the high jump either, but now I see that the 'Fosbury flop' is a satisfyingly elegant way of getting your body over a bar as efficiently as possible. If someone had explained to me back then that your centre of gravity *doesn't even have to go over the bar*, I'd have been much more interested.)

Another level at which people commonly reach their abstract limit is calculus, which involves a completely new and strange – and, frankly, somewhat sneaky – way of manipulating and reasoning with 'infinitesimally small' things. Some people get through rigorous calculus, but unfortunately reach their limit halfway through their undergraduate maths degree or in the middle of their PhD.

Rigorous calculus is something most people only meet if they do maths at university. People find it hard because it doesn't fit with their idea of what mathematics is – pinning things down and getting answers to things with great certainty.

Calculus at school usually consists of answering specific questions, such as 'If you draw the graph of $y = x^2$ and shade in the space under the curve from $x = 0$ up to $x = 2$, what the area that you have filled in?'

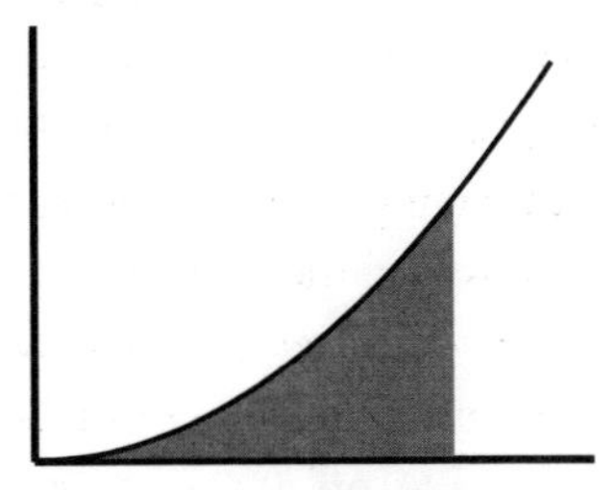

At school we are taught to answer this by 'integrating' x^2, which gives $\frac{1}{3}x^3$, and then evaluating this at $x = 2$, to give the answer $\frac{8}{3}$.

However, at university we *prove* that this argument is valid. At school you might see it justified somewhat experimentally, by drawing the graph on squared paper and then counting the squares under the graph. Some of the squares will only be partial squares, so you will only get a truly accurate answer if you use infinitesimally small squares.

Rigorous calculus makes this argument into something logically watertight, but baffles people because it doesn't pin down an answer in the way that people are expecting. Instead it says something like: there's no such thing as graph paper with infinitesimally small squares, so we use progressively smaller and smaller squares and observe that the answer gets closer and closer to $\frac{8}{3}$ as the squares get smaller. Then we prove that no matter how close we wanted it to get to $\frac{8}{3}$, there is a size of square that would get us that close.

A level at which advanced mathematicians sometimes reach their abstract limit is category theory. They react in much the same way that teenagers do when they meet *x*'s and *y*'s – they say they don't see the point, and resist any further abstraction. I am always reminded of Prof. John Baez, who said the following during an argument about abstraction on the worldwide 'Category Theory emailing list':

> *If you do not like abstraction, why are you in mathematics? Perhaps you should be in finance, where all the numbers have dollar signs in front of them.*

I haven't yet met my abstract limit, but I do remember various key moments in my life where I was pushing a boundary and felt I had to make a conscious effort to get over the next bar.

From numbers to pictures

My mother taught me how you can draw a graph of x^2, like this:

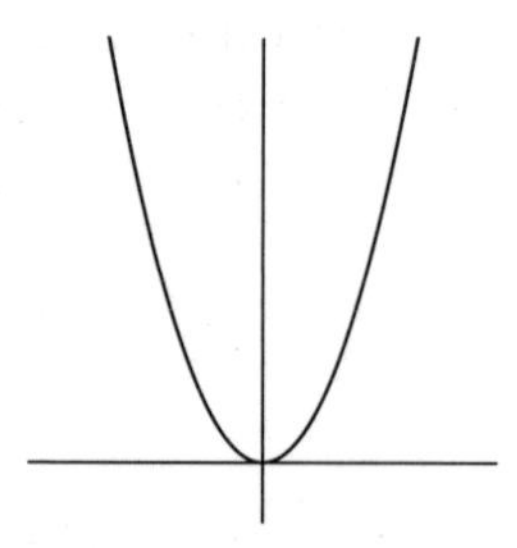

I distinctly remember my bafflement at the fact that you could turn the process of squaring numbers into a *picture* of a curve. I sat in our big green armchair at home thinking and thinking about this until my brain felt like it was popping out of my head. And in my memory this is the exact same feeling I've had every time I've thought about a difficult mathematical concept in my research.

From numbers to letters

I was perfectly comfortable solving equations with x's, say

$$2x + 3 = 7.$$

I knew this would turn into

$$\begin{aligned} 2x &= 7 - 3 \\ &= 4 \\ x &= \frac{4}{2} \\ &= 2. \end{aligned}$$

But then I met one with *a*'s, *b*'s and *c*'s instead of the numbers, something like

$$ax + b = c$$

and I vividly remember feeling completely at a loss as to how on earth to find out what x was in this case, without knowing a, b and c. I think I knew that I should start by subtracting b from both sides, but I had no idea what that would give on the right-hand side. I do remember that when someone explained to me that it would be $c - b$ I felt extremely stupid. Why couldn't I have worked that out myself? The answer is then

$$x = \frac{c - b}{a}.$$

Well, as I say to my students – feeling stupid for not having understood something before just shows that you are *now cleverer* than you were then.

From numbers to relationships

This is the last big leap of abstraction I remember having to make, and it was when I was first learning category theory. For the sake of completeness and perhaps amusement value, I'll include here what it was: it was the idea that a *one-object category is exactly a monoid*. Laugh as much as you like; there it is. I sat for days thinking about it and feeling like my brain was popping out of my head, just like when I was a child and thinking about a graph for the first time in my life. And the fact that a one-object category is exactly a monoid is now so obvious to me that I know I am definitely cleverer now than I was then. It's a bit early to explain this example now, but I'll come back to it in the second part of the book.

We will see that category theory studies relationships between objects. A *category* is a mathematical context for studying these relationships. A *monoid* is a mathematical context for studying something much more concrete: multiplication of things like numbers. The fact that a 'one-object category is a monoid' corresponds to viewing numbers as relationships between the world and itself. This sounds quite strange, but is remarkably powerful.

The goose that laid the golden eggs

Making machines for solving problems

It would be lovely to find a way of making golden eggs. But it would be even better to find a way of making a goose that lays golden eggs: a goose-that-lays-golden-eggs machine. But wouldn't it be even better to make a machine that makes these machines? A 'goose-that-lays-golden-eggs machine' machine. This is a form of abstraction. It's the idea of building a machine to do something, rather than directly doing the thing yourself. So really it's just a form of conservation of energy, or of reserving human brain power for the things machines can't do.

In order to build a machine to do something rather than doing it yourself, you have to understand that thing at a different level. It's like giving someone directions. When you walk somewhere you know well, you don't really think about exactly what streets you're walking on, or which way you're turning and when. You probably go somewhat instinctively. But when you're telling someone else how to get there you have to analyse how you do it more carefully, in order to explain it. You might have noticed that if you ask a local person where a certain street is, they will often not be very sure, as you don't really think about street names when you're wandering around your own town.

Something similar happens when learning a language. When you learn it yourself as your mother tongue, you don't really think about how it works – you pick it up from the adults around you instinctively. Then when you're an adult and a foreigner asks you to explain some aspect of the language that is confusing them, you have to go back and analyse how you speak in a completely different way.

If you're building a machine to make a cake, you'll have to analyse each step rather carefully in order to work out how to get a machine to do it. Even cracking an egg would require careful thought – how do we know how hard to tap the egg against the bowl?

The previous example of solving equations is an example of this type of machine. We start by understanding how to solve equations such as

$$2x + 3 = 7.$$

Then we make a 'machine' for solving all such equations, that is, we solve the equation

$$ax + b = c$$

because then a, b and c can be any numbers at all.

* * *

We can then try it for quadratic equations

$$ax^2 + bx + c = 0$$

and we learn that the 'machine' for solving these gives the famous solution

$$x = \frac{-b \pm \sqrt{b^2 - 4ac}}{2a}.$$

As a further level of building a machine that makes these machines, there is the *fundamental theorem of algebra* which tells us that every polynomial equation has at least one solution, as long as we allow complex numbers, which we'll come to later.

Cake cutting

An example of abstraction

I remember the first GCSE maths investigation I had to do at school. It was about cutting a cake into as many pieces as possible while making a fixed number of cuts. Obviously if you can only make one cut (in a straight line) you'll only get two pieces of cake, and if you can only make two cuts, you'll get at most four pieces. But what about three cuts? Four cuts? And so on?

The best answer for three cuts is: seven pieces of cake, like this.

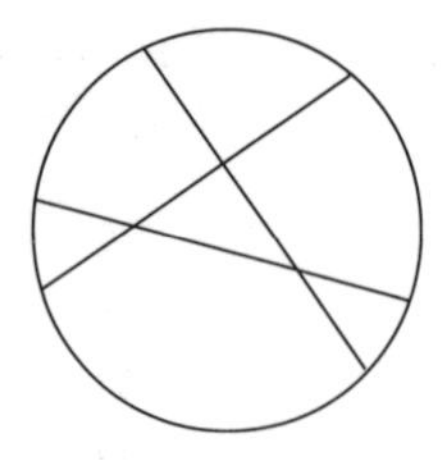

Your first thought about this might be the same as mine, which is: this is a stupid question, because who would ever cut a cake like that? You end up with pieces of all sorts of different sizes. What matters more in cutting a cake – efficiency or the sizes of your pieces of cake?

Setting aside the question of size for a second, the point of the investigation was to get us to try it experimentally for three cuts, four cuts, and so on, and then to get us to find a *formula* for the maximum possible number of pieces, in terms of the number of cuts you're allowed to make. That is, the aim is not just to solve the problem in any particular case, but to build a machine for solving the problem in *every* case. That is what a formula involving x's and y's and things really is – a machine. So you can feed in, say, the number of cuts you're allowed to make, and the machine will spew out the answer: the number of pieces of cake you get. A formula is even better than a machine: it actually tells you *how the machine works* – it's not just a mysterious black box. So if the formula says the answer is

$$\frac{x^2 + x + 2}{2}$$

this is a machine telling us that we can feed in the number of allowed cuts in the place of x, and the result will be the number of pieces of cake. This is a form of abstraction, because instead of dealing with actual problems, you're dealing with *hypothetical* problems. You're not solving the problem: you're solving the problem of solving the problem. Instead of writing the formula, you could make a table of answers like this:

No. of cuts	No. of pieces of cake
1	2
2	4
3	7
4	11
5	16
⋮	⋮

You can't make this table go on *forever* – it will have to stop somewhere, just because you'll run out of paper, not to mention years of your life. The formula, however, doesn't stop anywhere – it is a machine for giving you the answer for *any* number of cuts.

Perhaps you didn't have to do GCSE investigations, but perhaps you had children doing them and you were helping them. But you were trying to help them without actually doing the investigation for them. That is the meta-problem – instead of solving a problem, trying to solve the problem of getting someone else to solve the problem. Teaching is a bit like that, because you're not just telling people answers but trying to get them to find the answers: it's one level removed from answering the question yourself. Teaching teachers is another level of abstraction. And who teaches the people who teach teachers?

Making a cake isn't that clever, but inventing a new recipe for making cakes is a bit more clever. Discovering a new number wouldn't really count as 'interesting' because we already know the method for producing all new numbers. If you worked out how to cure cancer it would be somewhat immoral if you merely went round curing individual people's cancer instead of teaching the world how to cure cancer.

All of these examples of abstraction take us arguably one step further from reality, but have a broader scope as a result. If you shine a torch from further away, you will illuminate a larger area. But be careful not to shine it from *too* far away, as the light will then be too dim.

Abstract mathematics

Abstraction is the key to understanding what mathematics is. Abstraction is also at the heart of why mathematics can seem removed from 'real life'. That detachment from reality is where maths derives its strength, but also its limitations. Every level of abstraction takes it further from real life, and makes it harder to explain what the relevance to real life is, because the relevance comes from a domino effect – abstract mathematics might not be directly applicable to real life, but, rather, applicable to something else which is applicable to real life, or via an even longer chain of applications, for example:

Category Theory→Topology→Physics→Chemistry→Medicine

Abstraction is the key to understanding why mathematics is different from science at large. Evidence-based science proceeds with, obviously, evidence at its heart. You start with a 'hypothesis' – something you believe might be true, by general observation, by gut feeling, suspicion, anecdotes, or whatever. Now you need to test the hypothesis rigorously by finding evidence that holds up to scientific standards. These standards include things such as:

* You must have a sufficiently large sample size. Three or four cases is 'anecdotal' and could have been a fluke.
* The evidence must be controlled. You must be sure that you have accounted for other factors that might have affected the evidence, such as the placebo effect, socio-economic factors, the ages of people involved.
* The evidence must be unbiased. For example, with drug tests this means 'double blind' – the person taking the drug, and the person administering it, must not know whether it's a real drug or placebo.

In the end, the result is statistical. You come up with a large body of very convincing evidence, but your conclusion always has a percentage certainty attached to it.

Mathematics is different. The first step is the same – you start with a hypothesis that you think might be true for some reason. But instead of testing it rigorously using evidence, you test it rigorously using *logic*. The standard of 'rigour' is now completely different. It is nothing to do with sample sizes, because you don't actually use any samples – you only use thought processes. Bias doesn't come into it either, because all you're doing is applying rules of logic.

For example, to find out how much icing you need to cover a cake, you could do it experimentally – you could get a cake, ice it, and see how much you used. Or you could do it logically – you could do a calculation involving the surface area of a cake. To do this calculation you have to make an approximation of the shape of a cake – perhaps you assume it's perfectly round, and perfectly flat on top. Of course, no cake is ever *perfectly* circular and flat. But the advantage of this method is that you don't have to make any icing in order to find out how much icing you need.

Using logic instead of experiments has many different sorts of advantages.

Experiments can be impractical

Suppose you want to find out how many bricks you need to build a house. It's not very practical to build an entire house just to find out how many bricks you'll need. Or what if you want to work out how changing a road layout will affect traffic flow?

Experiments can be dangerous

What if you want to find out how much traffic a bridge can carry? You can't just get loads of traffic to drive across it and see when it collapses.

Experiments can be impossible

What if you're trying to work out why the sun rises every day, or why the planets behave the way they do? You can't just change the conditions of outer space and then see how the planets behave differently.

Experiments can be undesirable

Suppose you're trying to work out how an infectious disease can spread across the country. You can't just unleash the disease and see how it spreads, because that's exactly the thing you're trying to avoid.

Experiments can be immoral

At the time of writing, there is a suggestion that a badger cull will reduce instances of tuberculosis in cows. How can this be tested? Is it morally right to kill a whole lot of badgers to see what happens?

In all these cases, there is an important advantage to working theoretically rather than experimentally, an advantage to using logic rather than evidence. The final crucial advantage is that, with logic, the conclusion is not just 'almost certainly true': it is irrefutable.

How does logic work?

A logical argument is a series of statements, each of which follows from the previous one using only logic. That's all very well, but where does it start? You always have to start with a basic set of assumptions. For example, you might assume your cake is perfectly circular. You might assume that an infectious disease has a 50% chance of being passed from one person to another if they meet. These basic assumptions are part of the process of abstraction. They usually involve turning your real-life objects into something theoretical, so that you can reason with them using logic. The downside is that your theoretical situation won't be *exactly* the same as your real one. But the upside is that you can now apply logical process to work things out about them. The inaccuracy of your final answer will now come from the information you threw away when you performed the initial abstraction. This is very different from statistical results, where the inaccuracy of the final answer comes from a small percentage possibility that your hypothesis was wrong despite the evidence.

The *mathematical method* (as opposed to the more talked about 'scientific method') involves making very clear what your assumptions are. Then people can then disagree with your assumptions, but they aren't entitled to disagree with your overall conclusion, which is:

> *If* we make these assumptions, *then* this conclusion is true.

For example: if one chicken can feed ten people, then two chickens can feed twenty people. You can argue about how many people one chicken can really feed (probably not ten people unless it's a scary genetically modified giant chicken), but you can't argue with the fact that:

> *If* one chicken feeds ten people, *then* two chickens feed twenty.

But there's still a possible flaw here: are all the chickens the same size? We probably need to add an assumption saying 'All chickens are about the same size' to ensure that the situation behaves mathematically.

Is this an unrealistic assumption? If you're going to order whole roast chickens for a party with forty people, you're probably going to do a calculation somewhat like this, even though chickens aren't all *exactly* the same size. But on the other hand, you might proceed experimentally instead: you might rely on the experience of the caterer, who has probably held enough parties to have experimental evidence of how many chickens to get for forty people.

Abstraction can be difficult because it takes us out of the realm of physical objects, and into the realm of 'ideas' that we manipulate only in our head. But there are some abstract ideas we're so used to that we don't even notice how abstract they are any more. If we think about the size of an average chicken, that's an abstraction right there: an 'average chicken' isn't a real chicken we're considering, it's just an idea of a chicken. As I mentioned before, numbers are abstract. The numbers 1, 2, 3, 4, and so on, are only *ideas*. Because they are ideas, we can manipulate them just using logic.

The wonderful thing about abstraction is that when you get very used to an abstract idea, it starts to *feel* like an actual object instead of just being a made-up idea. You're probably quite comfortable with '2' as a concept. That means you're comfortable with that level of abstraction. Perhaps you're less comfortable with exactly what '−2' is. What about the square root of 2? It's a number such that when you multiply it by itself, the answer is 2. But what actually is it? You might think it's 1.414..., but that is a decimal that goes on forever without recurring – you can't write the whole thing down, so how do you know what it is? What about the square root of −1? We'll investigate these questions more later, and look at why rigorous mathematics has much more trouble with the square root of 2 than with -2 or even the square root of -1, although intuitively the square root of -1 is much harder to think about because nothing like it ever appears in 'real life'.

Part of the process of abstraction is like using your imagination. Mathematical abstraction takes us into an imaginary world where anything is possible as long as it's not contradictory. Can you imagine transparent Lego? That's not so difficult, but what about squashy Lego? That's a bit more strange. What about Lego that spontaneously changes colour when you touch it? Four-dimensional Lego? Invisible Lego? Lego that can make coffee for you in the morning? Obviously just because you can imagine something doesn't mean it actually exists in the real world – particularly if you have a very vivid imagination. The amazing thing about the world of maths is that mathematical things exist as soon as you imagine them. The more vivid your imagination, the more maths you have access to.

Another abstract concept that we're quite used to is shapes. What is a square? It's a shape with four equal sides and four equal angles. But are there actually any *perfect* squares in the world? No, any physical shape in the real world is not going to be an absolutely microscopically pedantically perfect square. Likewise circles. What about straight lines? Are there really any perfectly straight lines? Not really. And yet, we're comfortable with the idea of a straight line although the things in the real world are only approximations to this ideal.

Abstraction at work

Here I will give the abstract approach to the two example questions I posed earlier on, so you can see what it looks like.

> *My father is three times as old as I am now, but in ten years' time he will be twice as old as me. How old am I?*

I'll write x for my age, and y for my father's age. 'My father is three times as old as I am now' becomes

$$y = 3x.$$

So far so good. 'In ten years' time he will be twice as old as me' is a bit trickier. The key is that in ten years' time my age will be $x + 10$ and his age will be $y + 10$, and we know that his age will be twice mine at that point, so this turns into

$$y + 10 = 2(x + 10).$$

We can now substitute $3x$ into the second equation where y is, so we get

$$\begin{aligned} 3x + 10 &= 2(x + 10) \\ &= 2x + 20 && \text{multiplying out the bracket} \\ \text{so } x + 10 &= 20 && \text{subtracting } 2x \text{ from both sides} \\ \text{so } x &= 10 && \text{subtracting 10 from both sides.} \end{aligned}$$

So we can conclude that I am 10 years old (and my father is 30).

Note that we went through the following steps:

1. We started with a 'real-life' situation expressed in words.
2. We performed an *abstraction* to turn it into logical concepts.
3. We manipulated the abstract concepts, using logic.
4. We undid the abstraction, to get back to the real-life situation.

There's a further level of abstraction we can do here. The step we did helped us solve the problem stated in words above, but if we do another step, we can solve *all similar problems*.

In that problem we started with two specific equations:

$$\begin{aligned} y &= 3x \\ y + 10 &= 2(x + 10). \end{aligned}$$

But we can replace all those numbers by letters, so that we can solve any pairs of equations involving any numbers:

$$y = a_1x + b_1$$
$$y = a_2x + b_2.$$

The second equation of our original equations might not look like this to you, but when you rearrange it to get y by itself on the left, it turns into

$$y = 2x + 10.$$

Now we can solve the general pair of equations by equating the right-hand sides, since they're both equal to y on the left:

$$a_1x + b_1 = a_2x + b_2$$

and if we put all the x terms on one side, we get

$$a_1x - a_2x = b_2 - b_1$$
$$(a_1 - a_2)x = b_2 - b_1$$
$$x = \frac{b_2 - b_1}{a_1 - a_2}.$$

This last step is valid unless $a_1 = a_2$; in this case we are forced to have $b_1 = b_2$ as well, which means the two equations are the same, and we don't have enough information to pin down what x and y have to be – there will be infinitely many solutions.

Let's try the other example.

I have a recipe for icing the top and sides of a 6-inch cake. How much icing do I need for the top and sides of an 8-inch cake?

We *assume* that both cakes are round, and 2 inches deep. We need to find the area of icing used in the 6-inch cake, and the area used in the 8-inch cake, and see how much bigger the latter is. Because both cakes are round, we can save some effort by calculating the area of icing on a cake of radius r, and then we can use $r = 3$ or $r = 4$ afterwards (the radius being half the diameter).

* The top of the cake is a circle, so the area is πr^2.
* The side of the cake has an area which is the height times

the circumference. The circumference is $2\pi r$, so the area is $2 \times 2\pi r = 4\pi r$.

* Thus the total icing for radius r is $\pi r^2 + 4\pi r$.

We can now use this formula to work out the area covered by icing in each of the two cakes.

* For the 6-inch cake the radius is 3, so the total area covered by icing is

$$\begin{aligned}(\pi \times 3^2) + (4\pi \times 3) &= 9\pi + 12\pi \\ &= 21\pi.\end{aligned}$$

* For the 8-inch cake the radius is 4, so the total area covered by icing is

$$\begin{aligned}(\pi \times 4^2) + (4\pi \times 4) &= 16\pi + 16\pi \\ &= 32\pi.\end{aligned}$$

Finally we need to translate this into something we can use for our cake. We want to know how much to scale up the original recipe to make enough icing for the bigger cake, so we need to know how much bigger the second area is than the first. So we take the area we found for the 8-inch cake and divide it by the area we found for the 6-inch cake:

* The ratio of 8-inch icing to 6-inch icing is

$$\frac{32\pi}{21\pi} = \frac{32}{21}.$$

Now because this is only icing for a cake, and not something extremely critical like a dose of medicine, an approximate answer will do: $\frac{32}{21}$ is about 1.5, so you need to multiply your original recipe by 1.5 to have enough icing for the bigger cake.

The important thing to notice here is that we made an *assumption* that the cake is 2 inches high. So the final answer might be inaccurate, but only because of this assumption. So our final, irrefutable conclusion is

If all the cakes are 2 inches high,
then we need to multiply the original recipe by 1.5.

This cake example is somewhat more useful than the example with my father's age. Where the question of age was just a silly brainteaser, the question about icing was a genuine situation where the abstract thought processes helped us. We could have worked out the answer experimentally, by making a whole load of icing and seeing how much we needed for the bigger cake, but that would have been a waste of icing. The abstract approach used more brainpower, but wasted less icing.

3 PRINCIPLES

Conference chocolate pudding

Ingredients

2 large eggs

140 g caster sugar
140 g self raising flour
140 g butter softened

Cocoa powder to taste

About 7 squares of chocolate

Method

1. Cream the butter and the sugar.
2. Beat in the eggs, then fold in the flour.
3. Beat in cocoa powder until the mixture looks dark brown.
4. Half fill small 14 individual silicone cases with the mixture, then put half a square of chocolate in, and cover with more of the mixture.
5. Bake at 180°C for about 10 minutes. Eat immediately.

I call this 'conference pudding' because I first made it after a conference dinner when a whole group of mathematicians piled into my flat feeling merry and asked me to make pudding. It was a case of improvising something with whatever was in the kitchen. Fortunately, my kitchen always contains a large quantity of chocolate. Then I could follow some of the basic principles of cake making. Equal quantities of egg, flour, butter and sugar is a good basic starting point – other cake recipes can get very complicated, but what for? Chocolate usually makes people happy, and putting some in the middle of each pudding means that the middle is gooey, and the excitement of the oozing middle will distract people from whatever else happens with the puddings.

The point is that if you understand the *principle* behind a process rather than just memorising the process, you will be much more in control of the situation, better able to fix it when it goes wrong, in a better position to modify the process for different purposes, and better able to cope in extreme situations such as missing ingredients, broken equipment, drunkenness. . .

Drunk baking

Coping in extreme situations

Drunk driving is dangerous and to be avoided at all times. However, drunk baking is quite fun if you understand what you're doing. There are other reasons to understand the basic principles of cakes, instead of just faithfully following recipes. You might have friends with food intolerance – so you need to make cakes without wheat. (I've found that the best substitute flour for brownies is potato flour, for crumble it's oat flour, and for pastry it's rice flour.)

Perhaps you want to make cakes with less fat. So you need to understand the role the fat is playing in the cake – creating air bubbles – so that you can replace it with something that will play the same role, for example, curiously, apple purée.

Understanding the principles behind methods also helps you take shortcuts without ruining everything, and if you're lazy like me, you'll be looking for shortcuts all the time. Or simplifications because, for example, separating eggs turns out to be much harder when you're drunk. Recipes involving chocolate often say something like:

Break the chocolate into small pieces and place in a heatproof bowl, set over a pan of simmering water, ensuring that the base of the bowl does not touch the base of the pan. Stir until melted.

But what they really mean is 'Melt the chocolate'. I eventually became curious about this business of not letting the bowl touch the bottom of the pan and so tried it – and it didn't seem to make any difference. I also often melt chocolate in the microwave or, best of all, the oven at a low heat. Recipe books rarely explain why they're telling you to do something, which I find frustrating. But then, understanding is power, and if you help someone understand something you're giving them power. Perhaps those writers don't want us to understand too much, otherwise we wouldn't need them to invent recipes for us?

For a mathematical example, it's useful to memorise your times tables, rather than having to count up on your fingers each time. But it's also useful to understand how to work out the times tables, in case you forget, and need to work it out from scratch.

By the way, recipes always say to use cream of tartar in meringues, but I never have, and my meringues seem perfectly fine. Delicious, even.

Welding

My attempt at understanding how cars work

When I was 16, I appeared on television welding. I was working on a car project at school, where we were taking an old MG apart under the supervision of two of our physics teachers, and rebuilding it with new parts. For some reason I was the best at welding, and I also found it quite exciting – the noise, the sparks, the heat, the danger and the 'magic' of joining metal together using heat. By contrast I wasn't very good at understanding how the whole car worked. I just welded whatever I was told to weld.

I suppose the local TV station thought it was funny that a bunch of girls was building a car (I hope that doesn't seem so funny these days), so they turned up to film us one day, and I was duly filmed welding something.

The interviewers asked us if we were doing it to impress future boyfriends, but I was doing it because I wanted to understand the principle of how a car worked. I still think it's a good idea to know the principles of something that you're using all the time, so that you're less at its mercy when it goes wrong, and so that you have a better chance of getting the best out of it. The trouble is that, with the advance of technology, the workings of things have become more and more deeply embedded in electronics and code, so it's much harder just to take something apart and stare at it. When I did drive a car around for a while, most of the things that went wrong with it were electronic rather than mechanical.

Unfortunately my idea with the car project failed. I know how to weld, but I ended up none the wiser about how a car works, so now if my car breaks down I still have little choice but to take it to an expert. At least if my maths 'breaks down' I have a chance of fixing it myself – that is, I can check my reasoning and see where my logic was flawed.

Maths can be demoralising for children if they keep getting the wrong answer but they don't see what went wrong. That's why it's so important when teaching maths to understand the student's way of thinking, and point out what was wrong with their logic, not just what was wrong with their final answer.

Mars

What do we look for first when looking for life there?

When we look for the possibility of life on another planet, we start by looking for signs of water. This is because we've worked out, or decided, that water is pretty much crucial for making life viable.

When the explorers of Europe went and colonised faraway lands they did many things wrong (not least, perhaps, the colonisation in the first place). But one of the things they did wrong was try to bring crops with them from Europe to grow in lands with rather different climates. They had not in any way understood what was necessary to make those crops grow, and that the crops would therefore fail in those hotter, harsher lands. Or maybe they hadn't understood just how different the faraway climate was going to be. In any case, the crops failed.

One purpose of studying the principles behind things is to understand what is really making a situation work, so that you know whether it will still work when you go to a faraway land. That's a faraway *mathematical* land.

For example, one of the mathematical lands we feel most at home with is the 'natural numbers'. These are the numbers we use for counting: 1, 2, 3, 4, and so on, and they're called 'natural numbers' for a reason – they feel very natural. The trouble is, they're so natural we don't even *notice* the things we're using about them. It's like the fact that if you break your arm, you suddenly notice all sorts of things that are difficult, that you entirely took for granted when you had the use of both hands. We might not really notice when we particularly need to use both hands at once, and when one solo hand will do. Brushing your teeth seems like a one-handed activity, but how do you get the toothpaste on the toothbrush? Eating crisps seems one-handed, but how do you open the bag?

Likewise with the natural numbers. We take for granted that we can add and multiply, and that it doesn't matter what order we do it. So $8 + 4$ is the same as $4 + 8$, and we often use this when we're adding up – it's much easier to add a small number onto a big number, rather than adding on a big number to a small number. This makes an especially big difference to small children who are still adding up by counting on their fingers.

Summing $2 + 26$ will take a very long time if they start with 2 and count on 26, but if they start with 26 and count on 2 it will be quite quick – the difficulty for the teacher is in convincing them that they will still get the same answer.

Likewise, 6×4 is the same as 4×6, which is a good thing, because it means we only have to remember half our times tables. Personally I can only do 4×6 by thinking of it as 'six fours' and not 'four sixes'. Likewise I have to think of 8×6 as 'six eights'. But 8×7 I have to think of as 'seven eights'. Here's a whole grid of which times tables I do and don't know – perhaps you have something similar but different? Do you know 'eight sixes' or 'six eights' or both?

	2	3	4	5	6	7	8	9
2	✓	✓	✓	✓	✓	✓	✓	✓
3		✓	✓	✓	✓	✓	✓	✓
4		✓	✓			✓	✓	
5			✓			✓	✓	
6		✓	✓	✓	✓	✓	✓	
7		✓	✓		✓	✓	✓	
8						✓	✓	
9							✓	

In this grid I'm reading down first, and along afterwards. So I don't know 'five sixes' but I do know 'six fives'. I have no idea how my brain ended up dealing with times tables this way. Fortunately, turning the multiplication around gives the same answer, so I can *deduce* all my times tables, even if I don't actually *know* them all.

But what if we went to a different mathematical world in which these helpful facts weren't true? We would need to think very hard about what the knock-on effect would be. All sorts of things would start going wrong. Would we be able to solve equations any more? Would we be able to draw graphs? Would our standard techniques for *anything* work any more? We'll find out later in the book.

A more interesting principle of the natural numbers is to do with prime numbers. Remember, a prime number is one that is only divisible by 1 and itself (and 1 doesn't count as prime). So the first few prime numbers are

$$2,\ 3,\ 5,\ 7,\ 11,\ 13,\ \ldots$$

Now if I think of any number at all, there will be a unique way of writing it as a product of prime numbers. For example, $6 = 2 \times 3$, and there's *no other way* of multiplying prime numbers together to get 6, apart from changing the order of multiplication, which doesn't count as different; $24 = 2 \times 2 \times 2 \times 3$, and there's *no other way* of multiplying prime numbers together to get 24; and so on. This is a very important property of the natural numbers, but it *doesn't* hold on all mathematical planets.

This has created problems for mathematical explorers just as for those trying to plant crops in unfamiliar climates. For example, several attempts at proving Fermat's last theorem turned out to be wrong because people thought they were working on a planet where this prime factorisation property was true, when it fact it wasn't. They had devised a brilliant mission to Mars assuming there was water there.

Fermat's last theorem was famously stated by Pierre de Fermat in the margin of one of his books in 1637. It is about the equation

$$a^n + b^n = c^n$$

where a, b, c and n are positive integers. When $n = 2$, this is related to Pythagoras' theorem about the lengths of the sides of right-angled triangles: the square of the hypotenuse (the longest side) is equal to the sum of the squares of the other two sides. Most right-angled triangles are doomed to have edges that are not whole numbers. For example, if the shorter sides are 1 cm each, the hypotenuse will have to be $\sqrt{2}$ cm, which is not rational, let alone a whole number. However, there are some well-known right-angled triangles that do have whole-number sides, for example 3 : 4 : 5 and 5 : 12 : 13, satisfying the above equation:

$$3^2 + 4^2 = 5^2 \quad \text{and} \quad 5^2 + 12^2 = 13^2.$$

By contrast, for higher values of n, it is not possible to find integers a, b and c satisfying this equation: this is Fermat's last theorem, but it was not proved until 1995 when Andrew Wiles published a proof using very modern techniques from apparently unrelated fields of mathematics.

The principles of numbers

What are the basic principles of numbers? We're so used to them that we don't even notice them any more. Here are some facts about numbers that you probably take for granted.

* We can add numbers together.
* We can subtract numbers, but the answer might be negative.
* We can multiply numbers.
* We can divide numbers, but the answer might be a fraction.
* If we add zero to a number, it stays the same.
* If we multiply a number by 1, it stays the same.
* You can't divide by zero.
* If you add a number to something and then take it away again, you get back to where you started.
* If you multiply by a number and then divide by it again, you get back to where you started.
* When you're adding numbers up, it doesn't matter what order you do it.
* When you're multiplying numbers it doesn't matter what order you do it. But when you're mixing up $+, -, \times, \div$, it does matter.

* If you multiply anything by 0, you get 0.
* If you multiply anything by -1, you get the negative of what you started with.
* 'Minus minus is plus.'
* If you add something several times, it's the same as multiplying.

This is an awful lot of 'basic principles', so you might wonder if these can be reduced to a smaller number of 'extremely basic principles'. Like the fact that there is only one Brownie Guide law:

> *A Brownie Guide thinks of others before herself and does a good turn every day.*

The principles I've listed get harder and harder, broadly speaking, as you go down the list. When you're first learning about numbers it's quite hard to get your head around why it doesn't matter what order you add and what order you multiply. What about the fact that multiplying by 1 doesn't change anything? A recent study of primary school children showed that they got this wrong an alarming number of times. What about multiplication by zero? *Why* is that zero? Or worse, why do we get 'minus minus is plus'?

You might wonder where those principles came from in the first place: finding the basic principles behind something is called *axiomatisation*, which we'll come back to later. The idea in maths is that we take the basic principles of one world, say numbers, and see what *other* worlds obey those principles. You might be surprised to hear that the fact that multiplying by zero gives zero is *not* a basic principle: it's something we can prove from even more basic principles, as we'll see later in the book.

Things that obey the same principles as numbers are forced to be quite a lot like numbers, but they still don't have to be actual numbers. For example, *polynomials* look like this:

$$4x^2 + 3x + 2$$

They're not actually numbers, but they also obey these same principles.

If we drop the requirement that the order of multiplication doesn't matter, we get even more examples. *Matrices* look like this:

$$\begin{pmatrix} 1 & 0 \\ 3 & 2 \end{pmatrix}$$

and they obey all the above principles about numbers, apart from the one about the order of multiplication. We do have to be a bit careful exactly what this means, and we'll see how later on when we do some axiomatisation.

That's the whole point of understanding the principles – so that you can apply them to places that aren't quite the same as the ones you first thought of.

Questions for the curious

Try colouring in the following 2×2 grid. The rule is that each of the two colours has to appear exactly once in each row, and exactly once in each column. You should find that there's only one way of doing it.

red	blue
blue	

Solution: There are only two colours to choose from, so we can just try them: blue won't work so it has to be red.

Questions for the bold

Try this 3×3 one, with the same rules.

red	blue	green
blue		
green		

There should still be only one way of doing it.

Solution: Start with the middle square. It can't be blue, because then it would be next to the other blue squares. We could try red, but then the square to its right would have to be green, and that would be next to the top right green square, which is not allowed. So the middle square has to be green, and the one to its right has to be red, and the whole square has to look like this:

red	blue	green
blue	green	red
green	red	blue

The principle of each thing appearing exactly once in each row and once in each column is a bit like a simpler version of sudoku, and is called the *Latin square* property. It's a very important principle in maths, when studying *groups*, a branch of maths we'll come back to later.

Questions for the daring

What about this 4×4 one?

red	blue	green	black
blue			
green			
black			

There are now exactly four possible ways of doing it.

Solution:

red	blue	green	black
blue	red	black	green
green	black	red	blue
black	green	blue	red

red	blue	green	black
blue	red	black	green
green	black	blue	red
black	green	red	blue

red	blue	green	black
blue	green	black	red
green	black	red	blue
black	red	blue	green

red	blue	green	black
blue	black	red	green
green	red	black	blue
black	green	blue	red

This is a profound question in a subject called the *classification of finite groups*.

The last question here is: would you have found this easier or harder if those had been numbers instead of colours? It didn't actually matter that they were colours.

1	2	3	4
2			
3			
4			

What about letters?

a	b	c	d
b			
c			
d			

Changing to numbers or letters doesn't change the mathematics behind the question, which is about the patterns involved, regardless of how the squares are labelled.

4 PROCESS

Puff pastry

Ingredients

450 g strong white flour
450 g butter

Cold water

Pinch of salt

Method

...

There are many different ways of combining these simple ingredients, and most of them will not result in puff pastry. Making puff pastry is a long and precise process, involving repeated steps of chilling, rolling and folding to create the deliciously delicate and buttery layers that make puff pastry different from other kinds of pastry. Puff pastry has a reputation of being difficult to make because of this process. Shortcrust pastry is much easier – it uses the same ingredients (but less butter) and you can simply throw them in a food processor.

One of the wonderful features of maths is that, like with pastry, it can use quite simple ingredients to make very complicated situations. This can also make it rather offputting, like making puff pastry. Actually I don't think puff pastry is that difficult if you follow the instructions carefully. But even if you don't want to try doing it yourself, perhaps you can still enjoy the fact that such simple ingredients can turn into delicious puff pastry. Maths is about understanding processes and not just eating end results.

The New York Marathon

Not just about getting from A to B

In 2005 I ran the New York Marathon. I think this is a great achievement, so I boast about it whenever I can. In all honesty it's a bit far-fetched to say I 'ran' it – it would be more accurate to say I 'trotted' it. But I did make it from the beginning to the end, and there are photos to prove it.

The New York Marathon is different from some other marathons, say the Chicago Marathon, in that you do actually travel from one place to another place: you start on Staten Island and end up in Central Park. Whereas in Chicago you start in Grant Park and you end in . . . Grant Park. However, nobody thinks that simply getting from *A* to *B* is the whole point of running the Marathon – it's *how* you get to the end. If it were just about getting to the end, then everyone at the Chicago Marathon would just stand still.

When you tell people you've run a marathon, it's actually a bit like telling people you're a mathematician – some people think you're amazing and other people think you're mad. Why on earth does anyone do it?

The point is the journey itself, not just the arrival at the destination. Some journeys *are* simply about getting somewhere (for example, going to work in the morning). But other journeys are about a process of discovery or appreciation. It's easy to think of maths as a process of getting the right answer. And some maths is like that. But *category theory*, like the New York Marathon, is more about the journey, and what you see along the way. It's not about what you know, but *how* you know it. This is a much more nuanced question. If I ask you 'Do you know such-and-such fact?' the answer will be either yes or no. But if I ask you 'How do you know this fact?' the answer could be very long and complicated, and a lot more interesting than the sheer fact of whether you know it or not.

Pickpocket/putpocket

When it's not just about the end result

Suppose you have a ten-pound note in your pocket. Now, without you noticing, someone steals it. Also, and more strangely, someone else slips a ten-pound note *into* your pocket. At this point, you believe you have a ten-pound note in your pocket. But your reason for believing it is completely wrong. So are you right or not? Your conclusion is correct, but your reasoning is wrong.

This would count as the *wrong answer* in maths, because we're interested in the process of getting to the right answer, not just the answer itself.

Here's an example of incorrect reasoning leading to the correct answer.

$$\begin{aligned}\frac{4}{6}-\frac{1}{3}&=\frac{4-1}{6+3}\\&=\frac{3}{9}\\&=\frac{1}{3}.\end{aligned}$$

The final answer is correct but this is simply not the correct way to subtract fractions. One correct argument puts everything over the common denominator 6:

$$\begin{aligned}\frac{4}{6}-\frac{1}{3}&=\frac{4}{6}-\frac{2}{6}\\&=\frac{4-2}{6}\\&=\frac{2}{6}\\&=\frac{1}{3}.\end{aligned}$$

Delusion

When the end doesn't justify the means

If someone is happy, but you think they are happy for the wrong reasons, do you intervene? What if they're happy because they're drunk the entire time? What if they're happy because they're convinced they're God? What if they're happy because they're convinced that a God you don't believe in is looking after them?

Would you rather they were correct, but unhappy? Or to put it another way, does the end justify the means?

Maths is a world in which the end does *not* justify the means: quite the reverse. The means justifies the end; that's the whole reason it's there. It's called *mathematical proof,* and we'll see what that looks like shortly.

Two wrongs make a right

Why it's not all about getting the right answer

I have marked exam questions where students were asked to do some sort of calculation in many small steps. As it turned out, there were several steps where they were prone to making a plus/minus error, which could result in them getting the answer wrong by a factor of -1. So if the answer was supposed to be 100, they would get the answer -100 by mistake.

The trouble was, if they made *two* of these errors, the error would correct itself and they'd go back to getting 100. I seem to remember there were about six steps with the potential for making this mistake. So as long as they made the mistake an even number of times, they would still get the right answer. But they would have two, four or six mistakes in their reasoning.

In maths beyond the level of arithmetic and other school maths, the only reason you know you have the right answer is by checking that your process was correct. It's not like when you're trying to find the Eiffel Tower and you know when you've found it because everyone knows what the Eiffel Tower looks like. It's more like explorers in times gone by, who had no GPS and no maps, so the only way they could know where they were was by plotting their route very carefully.

Why? Why? Why?

Why small children have a point

If you've ever spent time with a three-year-old, you'll know that they never stop asking why. Ever.

'Why can't I have more dessert?'

Because you've had enough.

'Why?'

Because otherwise you'll have too much sugar and won't go to sleep.

'Why?'

Um, because your blood sugar levels will spike and your metabolic rate will suddenly go up and...

Unfortunately we suppress this instinct in children, possibly just because it gets rather tiresome after a while. Possibly because we quite quickly get to the point where we don't know the answer, and we don't like having to say 'I don't know'. Or we don't like reaching the end of our own understanding of things.

But this natural instinct in children is beautiful. It's the difference between *knowledge* and *understanding*. Sometimes they're just trying to pester the adults, or put off going to bed, but often I think they really are baffled by things and are trying to understand them better.

At the heart of maths is the desire to understand things rather than just know them. In some ways I just never stopped being that toddler who keeps asking 'Why?'. Maths is the most satisfying way I found of answering those 'Why?' questions. But then, inevitably, I started asking 'Why?' about maths itself, and that's where category theory comes in.

Mathematical proof

In mathematics the question 'Why?' is answered in the form of a *proof*. Proof in maths means something a bit stronger than the word in normal life. As we discussed in Chapter 2, it's not about gathering evidence, but about using logic.

For example, you might try to prove that all crows are black. You start looking for crows. The first one you see is black. The second one you see is black. The third one you see is black. You keep going. When do you decide you have enough evidence that all crows are black? After a hundred? A thousand? A million? There could still be one freak crow out there that is purple.

The thing is that crows don't really behave according to *logic*, so a logical proof would be quite difficult. You'd have to do something like find some irrefutable genetic cause of crows being black.

This is why, in mathematics, we focus entirely on things that *do* behave according to logic. The evidence gives us a hint of something we might sit down and try to prove using mathematical methods – but it could still be wrong. It happens plenty of times in research maths that you sit down to try and prove something that you think might be true because of some 'evidence', and the whole thing turns out to be completely false.

What if we try and prove:

All squares have four sides.

This is a bit silly – it's inherent in the definition of a square that it has four sides. (Is it inherent in the definition of a crow that it is black?) We need to try and prove something that isn't simply true by definition.

Let's try and prove:

Any number divisible by 6 is also divisible by 2.

We could start by looking for some evidence. Which numbers are divisible by 6? Well 12 is definitely divisible by 6, and yes, it's also divisible by 2. What about 18? Yes, that works. What about 24? Yes, that works. At this point you might *feel* very convinced. And that is important – feeling convinced is an important part of *being* convinced, and convincing people of things is the whole point of maths.

Can you instead work out *why* this is true? You might realise that it's something to do with 6 being an even number.

Let's try it for 24. We know that 24 is divisible by 6 because we know

$$24 = 6 \times 4.$$

But also

$$6 = 3 \times 2$$

so we can substitute this in, giving

$$24 = 3 \times 2 \times 4$$

which shows that 24 is divisible by 2. We could also split the 4 into its prime factors as well and get the prime factorisation of 24 that we saw in the previous chapter

$$24 = 3 \times 2 \times 2 \times 2$$

but we don't need to here – once one 2 has appeared in the product we know that 24 is divisible by 2 and we can stop.

Does this mean that any number that's divisible by 6 must also be an even number? In fact it does, and now we'll investigate why. First we should make that fact more precise by the following statement. We'll write A to stand for 'any number'.

If A is divisible by 6 and 6 is divisible by 2, then A is divisible by 2.

We can now make this work more generally and instead of 6 we could have any number *B*, and instead of 2 we could have any number *C*. Then we get the following fact:

If A is divisible by B and B is divisible by C, then A is divisible by C.

How do you feel about replacing all those numbers by letters? That is one moment that many people start feeling uncomfortable about maths. It's a step of abstraction too far for some people, but it has a point: we can now understand something more broadly true about numbers, not just about the particular numbers 6 and 2. Because *A*, *B* and *C* can now be *any numbers*.

Moreover, this process of taking a step back allows us to draw analogies with other things we might have seen. Can you see how the statement above, with *A*, *B* and *C*, is analogous to the following:

* If *A* is bigger than *B*, and *B* is bigger than *C*, then *A* is bigger than *C*.
* If *A* is cheaper than *B*, and *B* is cheaper than *C*, then *A* is cheaper than *C*.
* If *A* is equal to *B*, and *B* is equal to *C*, then *A* is equal to *C*.

This sort of relationship between *A*'s, *B*'s and *C*'s is called 'transitivity'. Mathematicians have given it a name because it crops up in many different situations, so it's useful to be able to refer to it quickly, and remind yourself of other similar situations. Here are some other relationships you can try this on.

Suppose *A*, *B* and *C* are people.

1. If *A* is older than *B*, and *B* is older than *C*, does that mean *A* is older than *C*?
2. If *A* is taller than *B*, and *B* is taller than *C*, does that mean *A* is taller than *C*?

3. If *A* is the mother of *B*, and *B* is the mother of *C*, does that mean *A* is the mother of *C*?
4. If *A* has the same birthday as *B*, and *B* has the same birthday as *C*, does that mean *A* is the sister of *C*?
5. If *A* is a friend of *B*, and *B* is a friend of *C*, does that mean *A* is a friend of *C*?
6. If *A* is married to *B*, and *B* is married to *C*, does that mean *A* is married to *C*?
7. Now suppose *A*, *B* and *C* are places. If *A* is east of *B*, and *B* is east of *C*, does that mean *A* is east of *C*?

The first two are definitely true. But the third one isn't – if *A* is the mother of *B*, and *B* is the mother of *C*, then *A* is the *grandmother* of *C*. So we say that being someone's mother is not transitive. Having the same birthday as someone is transitive, however. What about being someone's friend? Are you friends with all the friends of your friends?

What about being married to someone? If polygamy isn't allowed then you can only be married to one person. That means if *A* is married to *B*, and *B* is married to *C*, then *A* and *C* must be the *same person*. And that definitely means *A* is not married to *C*.

Finally let's think about (7). If the three places *A*, *B* and *C* are all within one city, or one country, then this is true. But if we encompass the entire world then we get into trouble because we can go round in circles. You can keep going east for a long time, and end up back where you started. This is a case where restricting your scope (to a single city or a single country) makes things easier to understand than looking at the entire world.[1]

Now let's go back to our example with the numbers. 'Being divisible by something' *is* transitive. But in order to prove that

---

[1] It might not sound like it, but that was actually a genuine mathematical example. Mathematicians study large and complicated surfaces by first restricted their scope to small neighbourhoods. They even use the word 'neighbourhood'.

properly, using rigorous logic, you have to turn 'being divisible by something' into a precise statement that can be manipulated using logic. This is another step that can make people feel uncomfortable. In order to get into a position to use logic, we have to leave the place where we use what we feel we understand about numbers – we have to leave our previous comfort zone. But the long term gains are large – there are places you can go with logic that you can't go with your gut feeling and instinct. It's like the fact that you have to leave the comfort of your home in order to get on a plane and see the world. Here's what that step looks like for our divisibility example.

A is divisible by B
means
A is a multiple of B
which means
$A = k \times B$ *for some whole number k.*

Now we're ready to go on our journey. When we do this in precise mathematical language we use a very specific structure so that everyone can agree on what just happened. It's like writing a story with a beginning, a middle and an end, except that you tell everyone what the end is going to be, before telling them what the middle is.

The beginning is where you state what your assumptions and definitions are. It's like setting the scene in a story, or writing out a cast list at the beginning of a play. It might look like this.

Definition 1. For any natural numbers A and B, we say *A is divisible* by B whenever $A = k \times B$ for some natural number k.

Now we tell everyone what the end is: the end result that we're going to aim for. In maths there's a hierarchy of names for 'end results' depending on how magnificent and ground-breaking they are supposed to be. A small one is called a 'lemma', a medium-sized one is a 'proposition' and a properly important one is called a 'theorem'. When something is suspected to be true but hasn't actually been proved yet, it's called a 'conjecture' or a 'hypothesis'. Hence there was the 'Poincaré conjecture' and the 'Riemann hypothesis' but there's also 'Fermat's last theorem'.

In fact the process of naming things is not very consistent: it's not clear why one thing should be a 'conjecture' and another a 'hypothesis'. Moreover, Fermat's last 'theorem', which we described in the chapter *Principles* was called a theorem for 358 years before a proof was ever published, which isn't really fair. Some extremely important things are called 'lemma', which sounds a bit like false modesty, but could also be because their importance was not recognised at first.

The Poincaré conjecture is about what sorts of three-dimensional shapes are possible. It is a three-dimensional generalisation of the following fact: if a two-dimensional surface has no edges, and is the surface of a three-dimensional solid with no holes, then it must be a sphere. It is hard to imagine what this could mean in the higher-dimensional version because it requires us to imagine four-dimensional solids: this is something that is difficult to visualise but easy to reason with in mathematics. Henri Poincaré proposed this conjecture in 1904, and it was called a 'conjecture' as Poincaré did not know how to prove it. It was finally proved by Grigori Perelman a hundred years later.

The Riemann hypothesis is about the distrubtion of prime numbers. A prime number is one that is only divisible by 1 and itself, and the first few are $2, 3, 5, 7, 11, 13, 17, \ldots$ You might think they form some sort of pattern, but they do not: there is no way of knowing where prime numbers are going to pop up. However, there are ways of predicting where they are more likely to appear, and the Riemann hypothesis gives a particularly good way of doing this. Bernhard Riemann proposed it in 1859 and to date it has still not been proved, so it is still called a 'hypothesis'.

The thing we're proving here about divisibility is fairly important in the context of numbers, so I'm going to call it a proposition.

Proposition 2. *If A is divisible by B and B is divisible by C, then A is divisible by C.*

Now that I've told you the beginning and the end of the story, I'm going to tell you the important part: the middle, the process of getting from the beginning to the end. This is called the proof.

Proof. Suppose:

1. A is divisible by B, so $A = k \times B$ for some natural number k, and
2. B is divisible by C, so $B = j \times C$ for some natural number j.

Then $A = k \times j \times C$, and $k \times j$ is a natural number.
Therefore $A = m \times C$ for some natural number m.
So by definition, A is divisible by C.
The end. □

Mathematicians don't really write 'The end' at the end; instead they will just draw that box □ to signify the end, or they will write 'QED' which stands for 'Quod erat demonstrandum', which roughly translates as 'which is the thing that we were supposed to demonstrate'.

Did you get lost somewhere in that proof? Were you perfectly happy with the original answer before we went into the mathematical details? Here are some other 'why' questions with various levels of answer. You can ask yourself whether you find each answer inadequate, satisfying or over the top, to see what sort of level of abstraction you like.

Question: Why does anyone use a three-legged stool?

a. Because a three-legged stool is more stable than a four-legged stool.

b. Because if you try and put four legs down on the floor, one of them might stick up a bit more than the others, leaving a gap between it and the floor, which means the stool could wobble.

c. Because given any three points in three-dimensional space, there is a plane that goes through them all. Whereas given any four points, there might not be a plane that goes through them all.

Question: Why does an octave sound nice whereas other combinations of notes sound discordant?

a. Because an octave basically consists of two versions of the same note, so they fit nicely with each other.

b. Because an octave is a natural harmonic, so when you play one note the harmonic of the octave above is already sounding anyway.

c. Because the wavelength of an octave above is exactly half the wavelength of the octave below, so there's no interference between them.

In each case all three answers are correct, but offer different levels of explanation. It is a matter of personal taste whether you are satisfied with the first answer or are still curious and seeking further explanation. It's about what sorts of facts you're happy to accept as 'basic' or 'given'. Maths tries to take almost nothing as basic or given, apart from the rules of logic. It always seeks further explanation.

5 GENERALISATION

Olive oil plum cake

Ingredients

2–4 plums

1 egg

100 g ground almonds

75 g agave or maple syrup

75 ml olive oil

Method

1. Slice the plums quite thinly and arrange them cut face down in a pretty pattern on a lined cake tin.
2. Whisk the rest of the ingredients together and pour gently into the tin over the plums.
3. Bake at 180°C for 20 minutes or until golden brown and set.
4. Turn out upside down so that the plums are on top.

If you've ever invented a new recipe, you might well have started with one from a book, or online, and modified it to your own tastes, whims or allergies. That is, you start with a situation you know and love, and see what you can do that's a bit similar but different – and maybe even better.

When I was little I was allergic to food colouring, so my parents lovingly worked out how to make jelly from scratch instead of from the appealingly (or appallingly) brightly coloured jelly packets. Later, I was going out with someone allergic to wheat, so I invented a lot of wheat-free desserts. (It's a bit easier to make wheat-free main courses.) Later on I started avoiding sugar, and I had other friends who were avoiding dairy... A modern complaint about cooking for friends is that so many people are following strange restrictive diets that it's impossible to cook for all of them at once. If you're faced with such friends you have several choices. You can refuse to invite them for dinner, you can ignore their dietary preferences and cook whatever you like, you can ask them to bring their own food, or you can rise to the challenge.

I invented the olive oil plum cake to be gluten-free, dairy-free, sugar-free and paleo-compatible. The only party guest who couldn't eat it was the one who was at that time only eating courgettes and ghee. Everyone said it was delicious, but when they asked me what it was I didn't know what to call it, because it's not really quite like a 'cake' – it's a *generalisation* of a cake. It has things in common with a cake, looks like a cake, is made like a cake, plays the role of a cake, but is still somehow not quite the same as a cake. It is useful in situations that an ordinary cake would not be able to handle.

This is the point of generalisation in mathematics as well – you start with a familiar situation, and you modify it a bit so that it can become useful in more situations. It's called a generalisation because it makes a concept more general, so that the notion of 'cake' can encompass some other things that aren't exactly cakes, but are close. It's not the same as a sweeping statement, which is a different use of the word, as we'll see later.

One example of generalisation is where we move from *congruent* triangles to *similar* triangles. Congruent triangles are ones that are exactly the same – they have the same angles and the same lengths of sides, that is, they're the same 'shape' and the same 'size'.

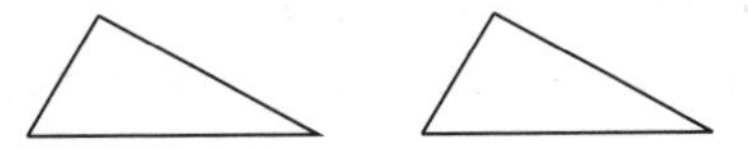

For *similar* triangles we only demand that they're the same shape, but not necessarily the same size – that is, they still have to have the same angles, but we drop the rule about having the same lengths of sides.

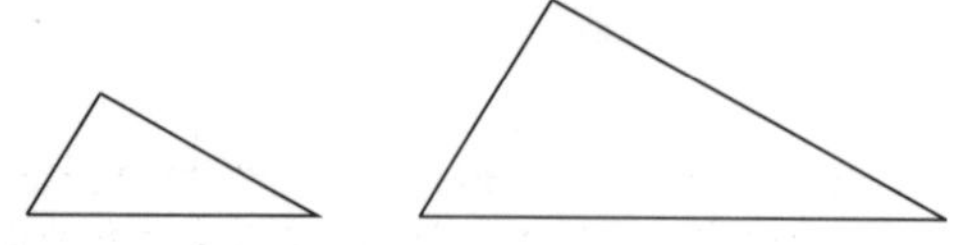

Because we've relaxed a rule, there are now more triangles that satisfy these conditions, but it still isn't total anarchy.

Flourless chocolate cake

Inventing things by omission

Imagine trying to 'prove' that you really need to boil water to make tea. You would probably just try to make tea without boiling the water. You discover that it tastes disgusting (or has no taste at all) and conclude that, yes, you do need to boil water to make tea.

Or you might try to 'prove' that you need petrol to make your car go. You try running it on an empty tank and discover it doesn't go anyway. So yes, you do need petrol to make your car go.

In maths this is called *proof by contradiction* – you do the opposite of what you're trying to prove, and show that something would go horribly wrong in that case, so you conclude that you were right all along.

Here's an example of a small proof by contradiction. Suppose n is a whole number and n^2 is odd. We're going to prove that n has to be odd as well.

We begin by assuming the opposite is true, so we suppose that n^2 is odd, but that n is even. However, an even number times an even number is always even, so this would make n^2 even. This contradicts the fact that n^2 was supposed to be odd, so we must have been wrong to assume the opposite. So the original statement that n is odd must be true.

Sometimes, proof by contradiction can be very unsatisfying, because it doesn't really explain why something *is* true, it just explains why something *can't be false*. We'll come back to this later when we talk about the difference between 'illuminating' and 'unilluminating' proofs, and the background assumption that if something isn't false then it must be true.

A famous, longer, proof by contradiction proves that $\sqrt{2}$ is *irrational*, which means that it can't be written as a fraction a/b, where a and b are integers (whole numbers). You might know that $\sqrt{2} = 1.4142135\ldots$ and that this decimal expansion 'goes on forever without repeating itself'. This is related to being irrational, but is not a proof. Here is a proof.

Proof. We start by assuming the *opposite* of what we are trying to prove, so we assume that there are actually two whole numbers a and b where $\sqrt{2} = \frac{a}{b}$. The trick is also to assume that this fraction is in its lowest terms, which means you can't divide the top and bottom by something to make a simpler fraction.

Now we square both sides to get

$$2 = \frac{a^2}{b^2}$$
$$\text{so} \quad 2b^2 = a^2.$$

So far so good. Now we know that a^2 is two times something, which means it is an *even* number. This means that a has to be an even number as well, because if a were odd then a^2 would also be odd.

What does it mean for a to be even? It means it is divisible by 2, which means that $\frac{a}{2}$ is still a whole number. Let's say

$$\frac{a}{2} = c$$
$$\text{so} \quad a = 2c$$

and now substitute that into the equation above, so we get

$$\begin{aligned} 2b^2 &= (2c)^2 \\ &= 4c^2 \\ \text{so} \quad b^2 &= 2c^2. \end{aligned}$$

Now we can do the same reasoning on b that we just did for a. We know b^2 is two times something, so it's even, which means b must be even.

Now we've discovered that a and b are both even. But right at the beginning we assumed that $\frac{a}{b}$ was a fraction *in its lowest terms*, which means that a and b *can't* both be even. This is a contradiction.

So it was wrong to assume $\sqrt{2} = \frac{a}{b}$ in the first place, which means that $\sqrt{2}$ cannot be written as a fraction, so is irrational. □

Proof by contradiction can be very efficient, and mathematicians sometimes use it as a last resort when they can't work out how to prove that something is true directly – they instead try to prove that it can't be false. Sometimes that kind of proof doesn't turn out the way you're expecting it to. Maybe you try to prove that you really need flour to make a chocolate cake. So you make it without flour . . . and you discover that it's really not that bad. In fact, you've invented a whole new kind of cake: the flourless chocolate cake, now popular in many fancy restaurants.

Likewise yeast and bread. You might try to prove (no pun intended) that you definitely need yeast to make bread. So you try making it without yeast – and you've 'invented' unleavened bread.

This can happen in mathematics as well – you set out to try and prove that something can't be done, and you accidentally discover that it actually can, although maybe something slightly different results. This is one way that generalisation can turn up, almost by accident. One of the most important examples of this is from geometry, involving parallel lines.

Parallel lines

The genius of Euclid

Once upon a time Euclid set out to write down the rules of geometry. The idea was to axiomatise geometry, that is, write down a short list of rules from which all geometrical facts could be deduced. The idea is that your basic rules should be absolutely fundamental, things so basic that you can't imagine deriving them from anything else – they simply *are true*.

Anyway, Euclid came up with four very simple and obvious sounding rules, and one annoyingly complicated one. They went something like this:

1. There's exactly one way to draw a straight line between any two points.
2. There's exactly one way to extend a finite straight line to an infinitely long one.
3. There's exactly one way to draw a circle with a given centre and radius.
4. All right angles are equal.

Those sound quite obvious, don't they? And then comes the fifth one:

5. If you draw three random straight lines they will make a triangle somewhere, if you draw them long enough, unless they meet each other at right angles.

The idea is that if your three straight lines meet each other at right angles, then two of them will be parallel, and no matter how long you draw them they will never meet up to form a triangle.

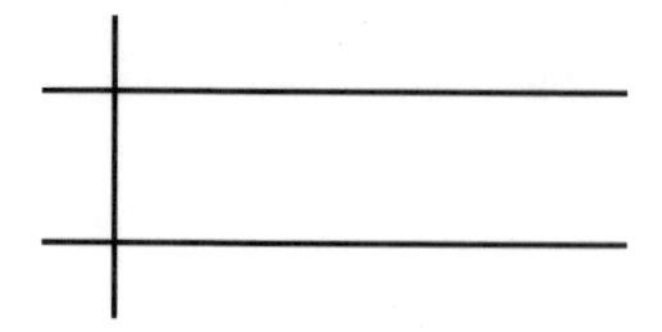

This is why the fifth law is called the 'parallel postulate', even though it doesn't explicitly mention parallel lines. The fifth law is also what tells us that the angles of a triangle always add up to 180°.

This last rule sounds so much more complicated than the others that people spent hundred of years trying to show it was redundant as a law, that is, that it could be deduced from the other four. Everyone knew they wanted it to be true; the only question was whether it needed to be enforced out loud, or whether it would automatically follow from the other laws, even if you didn't say it out loud.

People went round and round in loops and often thought they had proved it from the first four laws, when really they had accidentally used some assumption about geometry that seemed very obvious to them but was subtly equivalent to the fifth law. So, implicitly, they were *using* the fifth law to *prove* the fifth law – not that earth-shattering.

In the end people decided to try to prove it by contradiction, that is, they assumed that the first four rules held but that the parallel postulate did not, and then set about looking for things that would go horribly wrong elsewhere.

And the funny thing was, like with the flourless chocolate cake, nothing ever went wrong. It was just different – they had invented a new form of geometry.

We now know there are two types of geometry that don't satisfy the parallel postulate. There's the type where you imagine you're on the surface of something round like a sphere or a rugby ball. Here, the angles of a triangle add up to *more* than 180°. This is called *elliptical geometry*.

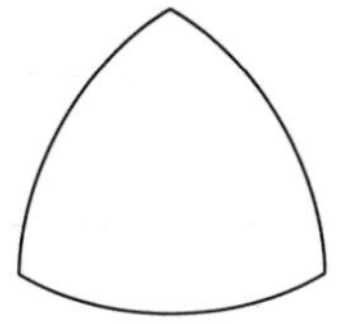

The other type is where you imagine you're on a surface curved the other way. Here the angles of a triangle add up to *less* than 180°. This is called *hyperbolic geometry*.

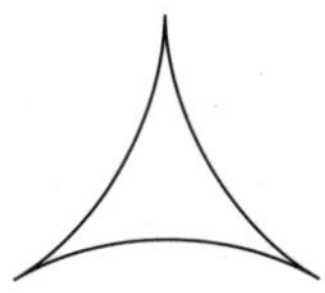

The original case where the parallel postulate does hold is like being on a flat surface, and is called *Euclidean geometry*.

Taxi cab

Generalising the notion of distance

We talk about distance 'as the crow flies', but when you're actually travelling it's unlikely you'll ever travel as the crow flies – so the distance from A to B will change depending on how you're travelling. How much you care about this will probably change too.

If you take a train, you usually buy your ticket at the beginning and then you don't worry about exactly how far the train is going. But if you take a taxi, it really matters how far the taxi is going. However, instead of the distance-as-the-crow-flies, we're thinking about the distance 'as the taxi drives'. The trouble is this can be affected by questions such as: is the taxi driver going the long way round? So we'd better assume we have an honest taxi driver, just like we assume the crow is going to take the shortest route rather than take some scenic detour. The important difference is that distance now depends on things like one-way systems, and suddenly the principles that are followed by crow-distance might not hold for taxi-distance. (Perhaps one day we'll have flying taxis that will really take us as the crow flies, but not yet.)

Here's an example. For a crow, the distance from A to B is the same as the distance from B to A. But this is not true for a taxi. For example, if you hail the cab at one end of a one-way street and get it to take you to the other end, that will be a much shorter journey than when you try to go home again and have to go the long way round.

If I get directions on Google Maps between Sheffield train station and Sheffield town hall I get this:

Station to town hall by car	1.4 miles
Town hall to station by car	0.9 miles
As the crow flies	0.5 miles

In a place like London it's quite hard to work out the taxi-distance from A to B, because the one-way system is complex, because the streets are so bendy, and because you're so concerned about how expensive the whole thing is becoming that you can't really focus on distances. So let's talk about Chicago, where it's much easier to work out taxi-distance for several reasons.

1. Mostly, it's a grid system, so the roads are all long and straight and meet at right angles.
2. The addresses are numbered according to distance, so '5734 South' (the number of the Maths Department of the University of Chicago) tells you how far south of zero the building is, not that it's the 5734th building down. This blew my mind when it was first explained to me. Since 800 = 1 mile, you can calculate how far your taxi has to go relatively easily.
3. The one-way system is fairly sensible, so that it's mostly possible to get where you're going without doubling back on yourself too much, as long as you know the system and make your turns at well-timed moments.

❹ Taxis are much cheaper than in London, so I don't get quite so worked up about how much it's going to cost.

Aside from getting worked up about the cost, this doesn't really depend on being in a taxi rather than any other kind of car. However, it is a genuine mathematical concept called the *taxicab metric*. It might be because it's the kind of thing mathematicians think about when sitting in a taxi, whereas if they're in a car one hopes they're concentrating on the traffic. We are gradually building up to the notion of 'metric', by investigating what sorts of properties distance-like notions should have.

Of course, Chicago isn't *precisely* a grid system at all times, and there are big highways that cut across the grid system at diagonals. So we're throwing away the details about diagonals for the time being. This process of throwing away inconvenient details is a form of 'idealisation' that is a key part of mathematics. This can seem frustrating (there simply *are* diagonal highways in Chicago) but the point is to shed light on something rather than to model it precisely. Our aim now is to shed light on the notion of 'distance'. Now that we've turned Chicago into an 'ideal grid' that taxis drive across, making only right-angled turns, the taxi-distance from A to B is simply

$$\text{horizontal distance} + \text{vertical distance}$$

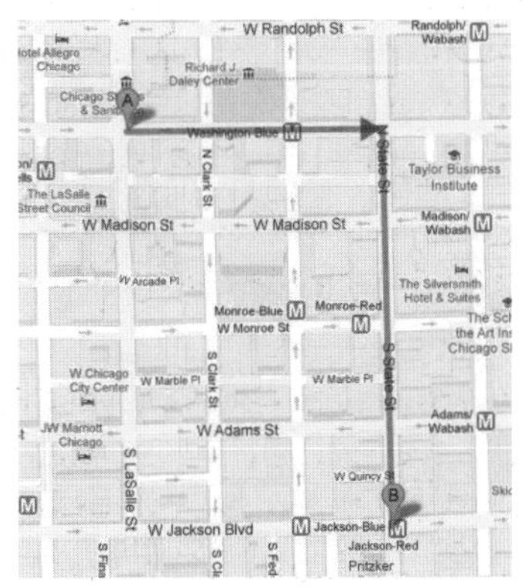

That is, no matter what clever route the taxi driver takes, it can't get any shorter than simply driving all the way across first, and then all the way down afterwards. Even if we make the turns in different places, say like this:

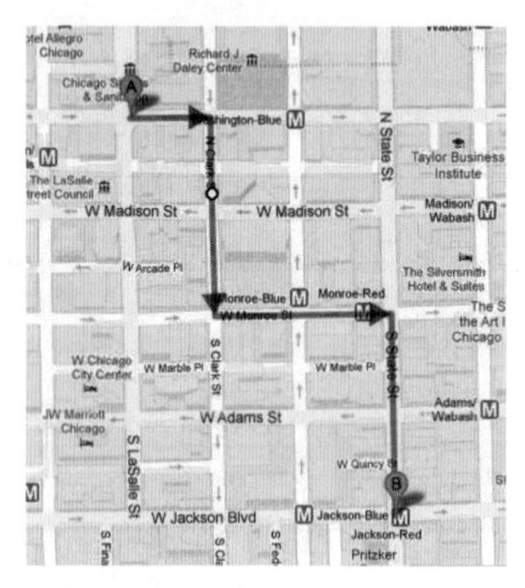

the distance is still the same, because we're not taking into account the time it takes to turn a corner. However, it would be longer if we did something really bizarre, like this:

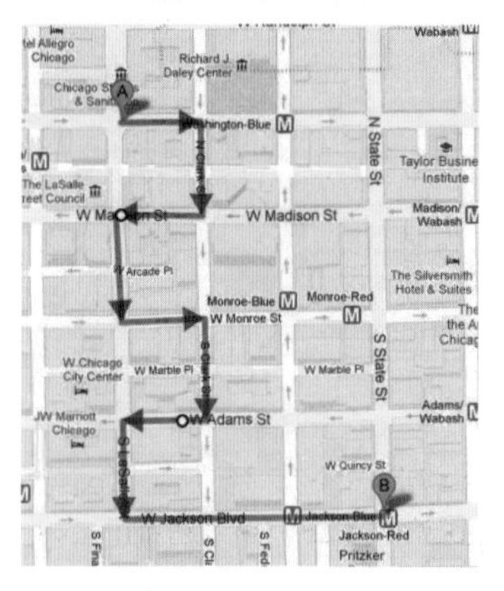

If you remember anything about Pythagoras' theorem, you may remember that it tells us how to calculate the length of the diagonal edge of a right-angled triangle. In our case, that's the distance as the crow flies.

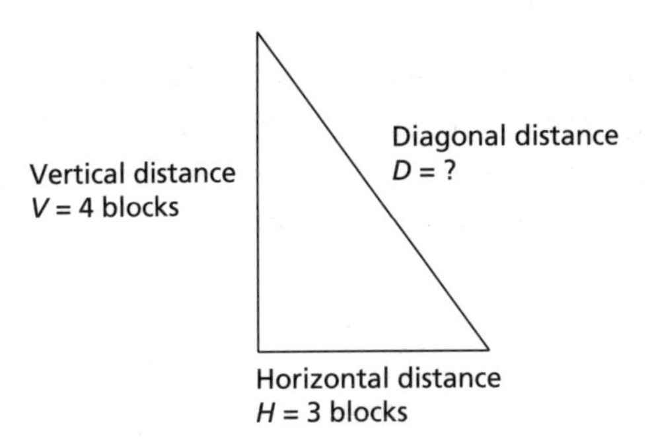

and in Pythagoras' case, that's called the 'hypotenuse'. Pythagoras' theorem says:

> *The square of the hypotenuse is equal to the sum of the squares of the other two sides.*

What this means on our diagram is:

$$D^2 = V^2 + H^2$$

and we can work out what the diagonal, crow-flying distance is:

$$\begin{aligned} D &= \sqrt{V^2 + H^2} \\ &= \sqrt{4^2 + 3^2} \\ &= \sqrt{16 + 9} \\ &= \sqrt{25} \\ &= 5. \end{aligned}$$

The crow only has to fly the distance of five blocks. However, the taxi has to go the vertical distance and the horizontal distance:

$$\begin{aligned} \text{Taxi distance} &= V + H \\ &= 4 + 3 \\ &= 7. \end{aligned}$$

The taxi has to drive the distance of seven blocks. The crow knows that taking some sort of diagonal route across the grid would definitely be shorter. But as a taxi, even if we tried to wiggle in a diagonal sort of fashion across the grid, it wouldn't help us – we'd still have to wiggle in only horizontal and vertical straight lines, and it would still add up to the same total horizontal distance and total vertical distance. And worse: we'd have to turn a lot of corners in the process.

Still, the taxi-distance is a perfectly good notion of 'distance', and is an example of generalisation. Again, we have taken a notion that we know and love, and we can now see what other notions are a bit like it but somehow different. What sorts of things should also count as 'distance'? This idealised taxi-distance obeys two crucial rules that crow-distance obeys:

1. The distance from A to A is zero, and that's the *only* way of getting a zero distance.
2. The distance from A to B is the same as the distance from B to A.

But there's also a third rule that is related to Pythagoras' triangle. It says that if you're trying to go from A to B, it can't be any better to go via some random other place C. Usually that will make it worse:

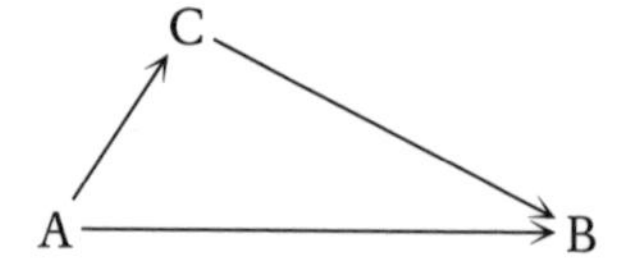

At best, C was on the way from A to B anyway, and going via C made no difference.

$$A \longrightarrow C \xrightarrow{\qquad\qquad} B$$

(You might have trouble trying to persuade a taxi driver of this though.) This rule about stopping off on the way is called the 'triangle inequality' because it's about the edges of a triangle –

not necessarily a right-angled one any more. It's like a very puny version of Pythagoras' theorem.

> **Pythagoras:** Yes! If we have a right-angled triangle then we can precisely work out the exact length of any side from the other two!
>
> **Triangle inequality:** Um, if we have a non-right-angled triangle then we know that the length of the third side will be *at worst* the sum of the other two.

Here 'worst' means 'longest' (because we're thinking taxis), so what we're saying is that if the sides of the triangle are x, y and z then the biggest x can be is $y + z$. You can imagine this as being an extremely long and thin triangle where the y and z edges have pretty much done the splits so that x has to be really long to accommodate them, like this:

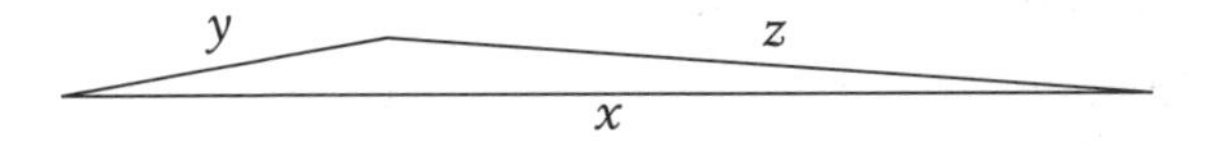

Now if we think of the edges of that triangle as the distances between our three places A, B and C then we get the 'intermediate stopping place' rule from before.

There are two curious things about this triangle inequality rule, I think. The first is that the taxi-distance still obeys this rule. The second is that there's a perfectly common 'distance-like' situation that does not, which is the cause of endless frustration to me: train tickets.

Train tickets

Generalising the notion of distance a bit more

If you've taken many trains around the UK you'll know exactly what I mean. It's the infuriating fact that sometimes, if you want to take a train from A to B, it's cheaper to buy two singles, via somewhere else. It's particularly stupid because you don't even have to take a different route – you just have to split the ticket in two. You don't always even have to get off the trains. Remember here we're not thinking about the actual distance covered in going from A to B, but the *cost* of going from A to B. In a sensible world, this would obey the triangle inequality – it would not cost less to go via some other place C. But in reality it does, or at least, it can.

For example, to go from Sheffield to Cardiff it can be cheaper to buy a single from Sheffield to Birmingham and a single from Birmingham to Cardiff.

To go from Sheffield to Gatwick it can be cheaper to buy a single from Sheffield to London and another from London to Gatwick.

To go from Sheffield to Bristol it can be cheaper to buy a single from Sheffield to Cheltenham and a single from Cheltenham to Bristol.

This is aside from the various other anomalies of UK train ticket prices, such as:

- Sometimes it's cheaper to go first class than standard.
- Sometimes it's cheaper to go further, for example London–Ely can be cheaper than London–Cambridge although the Ely train stops at Cambridge on the way.
- Sometimes it's cheaper to get a flexible ticket (where you can travel at any time of day) rather than one where you can only travel off-peak.

These last points are harder to explain in relation to the three rules of distance, because they're more to do with the interaction between cost and distance, or cost and time. So we'll leave those for now. Often in mathematics we focus on the easier things first, not because we're being wimps, but because the harder things are often built up from the easier things, and so we have to get the easier things right first.

In order to see why rules are imposed, it's often helpful to look at situations where they are *not* obeyed. Why is drinking alcohol not allowed on the Tube? Because it caused havoc. Why is smoking not allowed in Tube stations? Because there was a huge fire that killed people. This is similar to wanting to understand the principles behind things, rather than just memorising the rules or blindly following instructions in a recipe.

Now our three rules of distance are:

1. The distance from A to B is zero when A and B are the same place, and this is the *only* way the distance from A to B can be zero.
2. The distance from A to B is the same as the distance from B to A.
3. The distance from A to B can't be made any shorter by going via C.

Now that we've come up with a proposed list of axioms for the notion of distance, we'll do what is often the temptation when presented with a list of rules: we'll try to break them. The point of trying to break rules in mathematics is not to be arbitarily rebellious, but to test the strength and the boundaries of the world that we have set up.

We've seen distance-like situations that break rule 3 (train tickets) and 2 (one-way systems) but what about 1? You might think there's no real situation that violates rule 1 but here is one.

Online dating

Generalising the notion of distance yet further

GPS is marvellous technology. It means I get lost a lot less than I used to, especially on buses, where I can follow my position along the map on my phone, and then miraculously get off the bus in the right place.

GPS has also made online dating rather immediate. In the old, slow model, you could see if someone lived in the same city as you, or within say 100 miles, or 200 miles. With GPS, you can see how many *metres* away this person is *right now*. I've watched friends of mine do this in bars (just for a laugh, of course . . .) and the excitement of seeing how close someone is is palpable, especially when they're getting closer. 'Ooh, this one is only 200 metres away . . . 150 metres away . . . 50 metres away – wait, doesn't that mean he's in here?'

However, this can cause great disappointment because the distances are based only on GPS and don't take into account how far off the ground you are. A friend of mine was lonely in a hotel room somewhere and was perplexed at the number of interested parties who were supposedly 'zero metres away'. 'And yet,' he lamented, 'here I am alone in my hotel room.'

This is an example of a distance-like notion that does not obey the first rule of distance – that you can only be zero distance away if you're actually in the same place. This is relevant to some slightly more useful situations than lamenting your online-dating problems as well. For example, if your 'distance-like notion' is not actually the distance from A to B, but the amount of energy you need to expend to transport something from A to B. Then if A is directly above B you can just drop it, so the energy used getting it from A to B is zero, even though A and B are not in the same place.

A 'distance-like notion' is called a *metric* in mathematics. There's one more rule it has to satisfy that we didn't bother mentioning: that the distance from A to B is never negative. There are even situations where it's useful to relax this rule, for example, if we're studying how much it will cost to transport something from A to B. Not only might it cost you nothing (so the 'distance' would be zero), but someone might even pay you to do it. Coffee growers in Costa Rica are *paid* to send their coffee to Europe to be decaffeinated, because the caffeine that is extracted is so valuable to the makers of energy drinks.

Relaxing one or more of the usual rules for metrics is one way to generalise the notion of distance in mathematics. Another way combines generalisation with abstraction, and gives us the notion of *topology*, which we'll look at later in this chapter.

Three-dimensional pen

Generalising by adding dimensions

The problem with using GPS for online dating, as we saw above, is that it assumes we're only in a two-dimensional world. This usually works fine for finding your way around in a car, but not for finding a potential date inside a skyscraper, where the third dimension is rather important.

Increasing the number of dimensions is an important form of mathematical generalisation. There's a joke that if you're at a maths research seminar you can ask an intelligent-sounding question even if you don't understand anything, which is 'Can this be generalised to higher dimensions?'

A sphere is a higher-dimensional generalisation of a circle, if you think about a circle in the right way. Let's think about drawing a circle with a pair of compasses (although these days we all just draw circles by selecting a circle function on a computer). With compasses, you first choose a size (radius) for your circle, so let's say you open the compasses to 5 cm. Next you fix the pointy tip on the page where the centre of your circle will be, and then with the drawing end you essentially mark every point on the page that's exactly 5 cm from the centre.

Now imagine you have a pen that can draw in mid-air, which is something I've always dreamt of. Then you could fix your compass point somewhere, and use your mid-air pen to mark every point in the air that was exactly 5 cm from your chosen centre, *in all directions*. This would be a sphere.

At this point mathematicians are perfectly happy to generalise this to four, five, or even more dimensions, although we don't exactly know what that means. A sphere of radius 5 cm in four-dimensional space is 'all the points in that space that are exactly 5 cm from a fixed centre'. Because it's an *idea* rather than a physical object, it doesn't matter that we don't know what it looks like. It only matters that the idea makes sense. But just because one generalisation makes sense doesn't mean there aren't others that make sense, too.

Doughnut

A different generalisation of a circle

Imagine a doughnut. A ring doughnut.

When mathematicians say 'doughnut' they always mean a ring doughnut, at least when they're talking about maths. Perhaps they should start saying 'bagel' instead.

How would you generalise a bagel? The most obvious way is to give it more holes. A two-holed bagel!

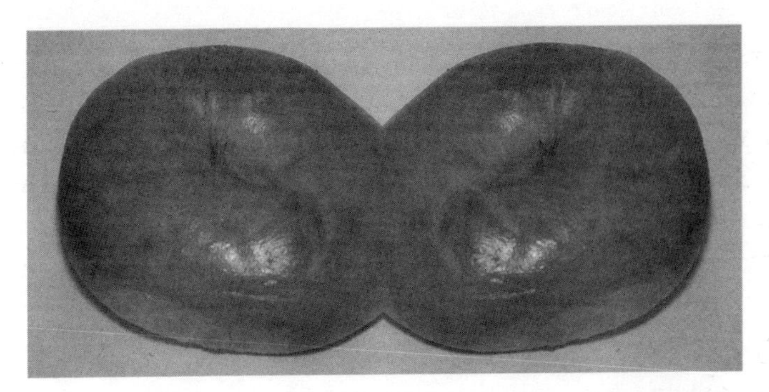

But there is another way to generalise it. For this we have to be a bit more careful about this bagel/doughnut of ours. When mathematicians think about doughnuts they're usually only thinking about the *surface* of the doughnut, not the solid doughnut. Just like when they say 'sphere' they only mean the surface of the ball, like the skin of an orange, not the whole orange. A sphere is like a balloon, with empty space on the inside.

Likewise for doughnuts. Perhaps you can imagine taking a toilet roll, magically turning it into stretchy rubber, and bending it round into a little hoop. Or perhaps imagine taking a Slinky and bending it round so that the ends meet up. It will look like a ring doughnut, but be hollow.

This is technically called a *torus*.

Now, let's think about how we made it from a toilet roll. You could also imagine trying to make it out of bubbles – the kind that come in a big bottle, with a big hoop that you can drag through the air to make bubbles, instead of blowing. Now imagine taking this hoop and dragging it through the air for a while – you make a sort of bubble tube as you go. Now imagine dragging it in a big circle so that it comes back to meet itself. It will be like a doughnut – a hollow doughnut. A hollow bubble doughnut.

We made this by dragging a hoop through the air in a circle, which shows that the torus is a generalisation of a circle – all we've done is draw in the air with a hoop instead of a mid-air pen. Now for the generalisation of the torus things are going to get a bit weird. Imagine dragging *an entire doughnut* through the air in a circle. It's pretty difficult to imagine what this looks like, because it doesn't really fit into three-dimensional space, but perhaps you can at least imagine that it's definitely not the same as a two-holed doughnut.

Sweeping statements

A different kind of generalisation

'It always rains in England.'

'The trains never run on time.'

'Opera is really expensive.'

'You always say that.'

These are all *sweeping statements*, or generalisations. But this is a different kind of generalisation from the kind where you turn a bagel into a two-holed bagel. This kind is not about relaxing conditions to allow more people in, but is more like ignoring outlying cases temporarily, to focus on the central part of the bell curve.

Of course, these sweeping statements aren't *entirely* true.

Occasionally, trains do run on time. And sometimes it stops raining in England. And you can easily get opera tickets in London for under ten pounds. And you don't really say 'that' (whatever that is) all the time, just under certain situations. The question is, do these exceptions matter? Do we study exceptions or do we study the main body of behaviour?

The answer, surely, is both. We can't really study one without studying the other. There are interesting things to be learnt from the extremities of behaviour, even if those extremities are rare, and so not at all representative. But how can we know in what way something is unusual if we don't also study what is usual? That involves temporarily ignoring the extremities.

Bagels, doughnuts and coffee cups

An introduction to topology

Combining our previous discussions about distances and bagels brings us to a branch of mathematics called 'topology', which studies the shapes of things. We've already seen ways of generalising the notion of 'distance' so that we have something a bit like distance, but not necessarily satisfying all of the usual rules that distance does.

But now we can generalise this even more, because there are times when we don't mind so much exactly how far apart two things are, but only whether we can get from one point to the other, and how. If you live in the south of England, the Isle of Wight is probably closer than Scotland, but the fact is that you can't just drive there – so it's a whole different kind of hassle.

Something similar can happen with neighbourhoods of a city. Some cities, like Chicago, can change rather abruptly from one block to the next, where one 'neighbourhood' ends and another begins. It doesn't matter that you've only travelled one street over – the distance is very small, but you've gone into a completely different neighbourhood.

When we don't care about distance it means we also don't care about size, just like with the similar triangles, so all these are 'the same':

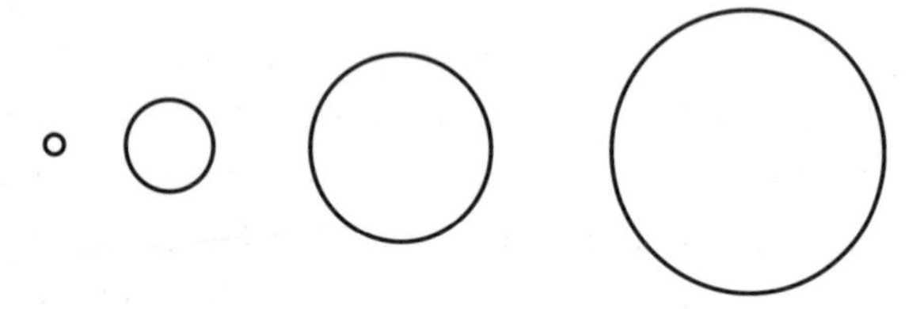

Another related thing we might not worry about is *curvature*, so these two shapes also count as 'the same':

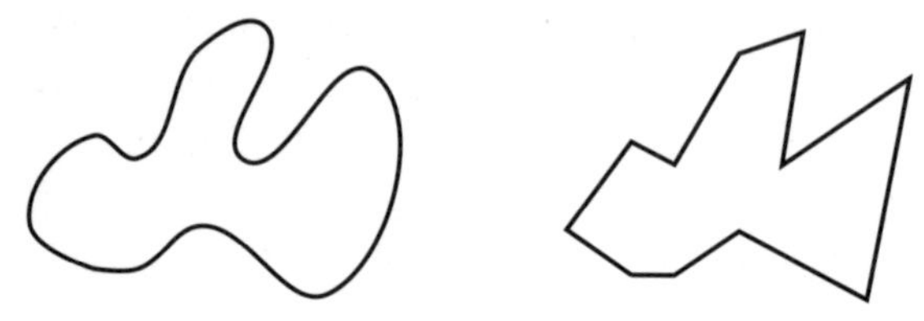

In fact the only thing we're really worried about is the number of holes something has. So now we have a system under which not only are all triangles 'the same', but triangles are also 'the same' as squares and circles: they're all shapes with one hole. However, a figure 8 is 'different' because it has two holes.

One way to think about this is to imagine that everything is made of plasticine or playdough, and you want to know if you can bend one shape into another without making any new holes or sticking anything together.

Question: Which capital letters of the alphabet are 'the same' in this bendy sense?

* There are letters with no holes: C E F G H I J K L M N S T U V W X Y Z.
* There are letters with one hole: A D O P Q R.
* There is just one letter with two holes: B.

What this says is that *topologically* almost all letters are the same. This is one of the reasons that computer recognition of handwriting is so hard.

We can also try this in higher dimensions. Imagine trying to make a bagel (a solid one, not a hollow one) out of a lump of playdough. There are basically two ways of doing it: you could either make a sausage shape and stick the ends together, or you could poke a hole in the lump. Either way, you've done something that shows that a bagel is not topologically the same as a plain lump. However, once you have your bagel/doughnut shape, you can make a coffee cup *without* making any new holes or sticking anything together. The doughnut's hole can turn into the handle of the coffee cup, and then you just need to squash an indentation in the rest of it to make the cup part. What this says is:

Topologically, a bagel is the same as a coffee cup.

However, the 'two-holed bagel' pictured earlier, is definitely different. The study of which things are topologically the same and different has many applications. For example, we talked about the mathematics of knots earlier on, and these are studied using topology. The amazing idea here is like the kind of drawing where instead of drawing on a blank page, you colour in an entire page and then erase parts to make a picture in white. Now we'll imagine doing this in three dimensions.

Imagine your mid-air pen again, and imagine that you have 'coloured in' the whole inside space of a box. Now you take a 'mid-air eraser' and erase a knot from what you coloured in. What is left is something with a curious shape that's almost impossible to imagine, but very handy to study mathematically.

A challenge for your imagination

The process of erasing something in three dimensions that we just described is called taking the 'complement'. Once we've done it, we can imagine that we are allowed to squash what's left just as if it were playdough, again without making any new holes or sticking things together. Can you imagine the following complements?

* The complement of a circle ○ is topologically the same as a sphere with a bar stuck across the middle of its empty insides:

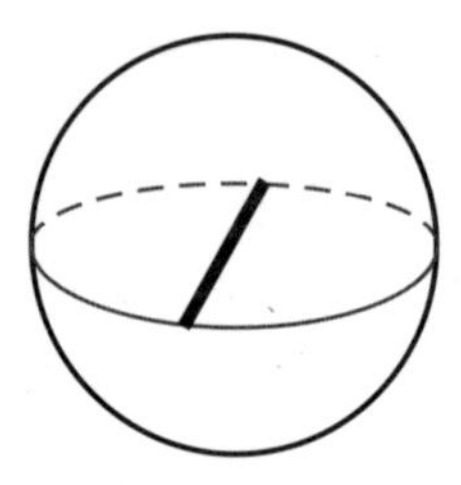

* The complement of two interlocking circles

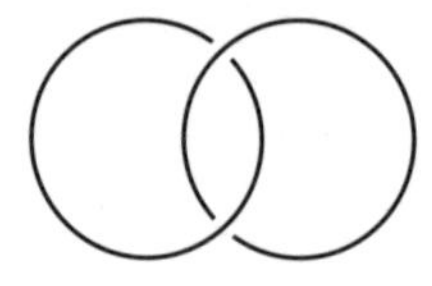

is topologically the same as a sphere with a torus stuck on the inside of the surface, in the empty space:

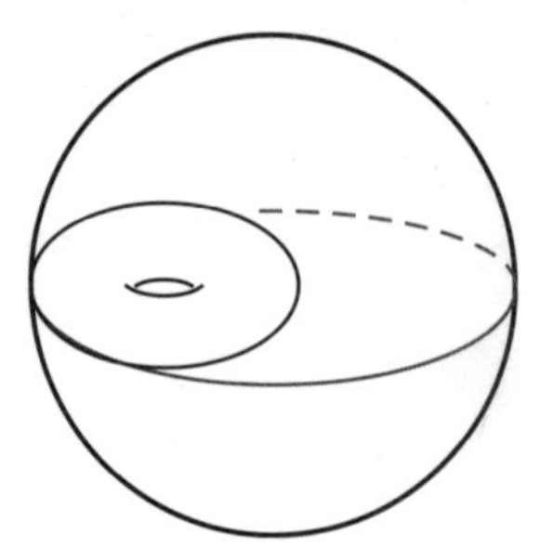

Those were only very simple shapes, and already it's very hard to imagine them in your head. The power of mathematics is that it enables us to study these things rigorously without having to imagine them at all.

Here's another example, involving cutting out shapes and sticking the sides together to make something three-dimensional. You may remember how to make a cube starting from a flat shape.

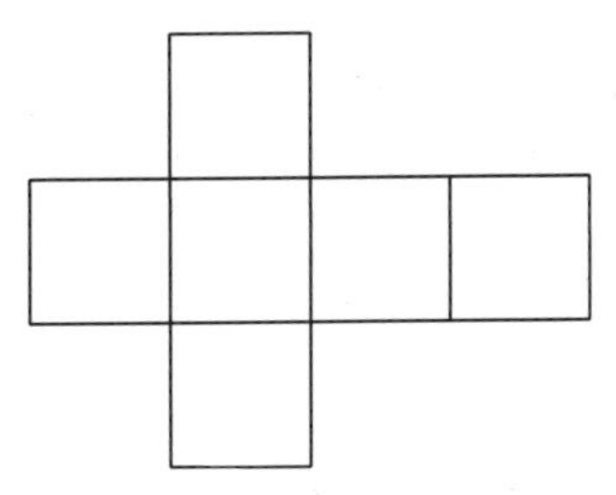

If you cut this out and fold it along the lines, you can stick the edges together to make a cube. If you try it with this one:

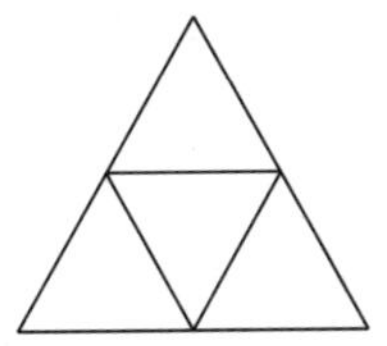

you will get a triangular pyramid which is technically called a *tetrahedron*.

Now imagine that you're actually making this out of bendy playdough paper. Now we can make a bagel/doughnut/torus out of a square like this – here we have to make sure we stick the edges labelled A to one another, with the arrows matching up, and likewise the edges labelled B:

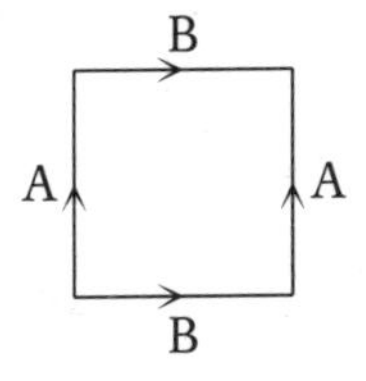

Now here's a serious challenge. Can you imagine what shape you'll get if you cut out this octagon and stick it together according to the labels?

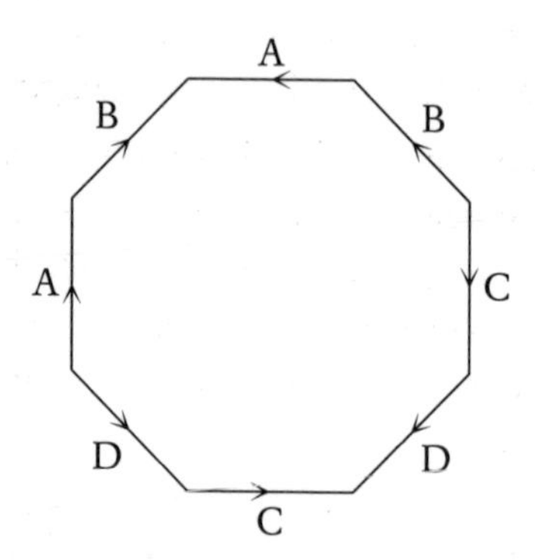

The answer is: a *two-holed bagel.*

Now imagine trying to generalise this for even more holes – it's pretty hopeless to try and do this in your head, but topology gives us a way of studying these things rigorously, for shapes much harder than those that our imaginations can ever visualise.

A generalisation game

What do the following shapes have in common?

square, trapezium, rhombus, quadrilateral, parallelogram

The answer is that they all have four sides. Now can you see how to arrange them in order of *increasing* generality? And what is the process of generalisation to go from each one to the next?

The answer is:

square, rhombus, parallelogram, trapezium, quadrilateral

The processes of generalisation are like this:

* A square has all four sides the same length, and all four angles the same.

* A rhombus only has all four sides the same, so the step of generalisation is to allow the angles to be different. However, they will be forced to be in pairs – the angles opposite one another have to be the same as one another just because the sides are all the same length.
* A parallelogram is like a rhombus but now only the sides opposite one another have to be the same length. There's no generalisation regarding the angles, which will still be forced to be the same in opposite pairs. Note that opposite sides are forced to be parallel, because of the opposite angles being the same.
* A trapezium only has the condition that one pair of opposite sides has to be parallel. So there's no longer any condition on the lengths of the sides or the sizes of the angles. They can now all be different.
* A quadrilateral is any old shape with four sides, so in this step we have generalised by dropping the condition that one pair of sides has to be parallel.

In this example we can see that each step of generalisation occurred by dropping some conditions on the shape in question, so that more shapes were allowed into the picture. Relaxing conditions slightly is one of the common ways of performing a generalisation in maths.

You might have noticed that there's another possible step in this generalisation, via another type of four-sided shape that we didn't mention above: the rectangle. A rectangle is a different way of generalising a square – where a rhombus still has the same lengths of sides, but possibly different angles, a rectangle has the same angles, but possibly different lengths of sides. When we relax rules one by one, we get different routes to generalisation depending on the order in which we relax the rules. Generalisation is not an automatic process. There are always different possible generalisations depending not just on how far you go, but on what point of view you take. This is one of the reasons mathematics as a subject keeps growing at an ever-increasing rate, as each generalisation gives rise to a multitude of others.

6 INTERNAL VS EXTERNAL

Chocolate and prune bread-and-butter pudding

Ingredients

250 g stale bread, without crusts

350 g chopped prunes

100 g dark chocolate

2 eggs

75 g caster and dark muscovado sugar in total

50 g melted butter

300 ml milk

Method

1. Break the bread into small pieces and make into bread crumbs in a food processor.
2. Beat the eggs and the sugar.
3. Melt the chocolate gently with the milk, and mix it into the eggs.
4. Pour over the bread and prunes in a large bowl, and leave it to soak for a few hours.
5. Mix in the melted butter.
6. Bake in a lined 8-inch square cake tin, at 180°C for 45 minutes or until set and slightly crispy on top.
7. Serve warm with chocolate sauce or chocolate custard.

This recipe for chocolate bread-and-butter pudding is something I came up with after making Christmas pudding one year. I had leftover bread (which I don't usually eat, and which had gone stale because I'd cut the crusts off) and prunes (which quickly go rock hard once you've opened the packet). And, of course, I always have plenty of chocolate in the house.

There are many dishes invented by our more frugal ancestors for using up leftovers. Cottage pie and shepherd's pie for using up leftover roast meat from Sunday lunch. Bread-and-butter pudding and French toast, or as the French call it *pain perdu*, literally 'lost (or wasted) bread', make use of stale bread by softening it up in egg and milk. There's the Chinese version, egg fried rice, where leftover rice is similarly fried with egg to soften it up again. Black bananas can be made into delicious banana cake. And everyone has their favourite dish to make out of the mountains of leftover turkey that are somewhat inevitable at Christmas. Curry? Pie? My favourite was my mother's turkey spaghetti salad with peanut sauce.

In all these cases it's sort of the wrong way round if you go and and deliberately look for the ingredients to make a dish that was supposed to be there to use up leftovers. Something similar can happen even if you're deliberately making a dish from new ingredients, as we mentioned in Chapter 1: you could pick a recipe and go shopping for the ingredients you need, or you could buy some ingredients that look interesting and invent something with them.

All this illustrates the difference between what I call *internal* and *external* motivation. If you set out with a recipe in mind, this is an external motivation. If you make something up from the ingredients you have, it's an internal motivation. Sometimes you set out with something in mind, but make it up as you go along, to see what will happen. If it then matches up with whatever you had in mind to make, your internal and external motivations have gloriously come together. Sometimes things turn out completely differently from how you were expecting them but are still fantastic. Or maybe you had no idea what to expect at all (like when I first tried making raw chocolate energy bars) but it's fantastic anyway. This is what we might call a 'happy accident'. That is different from the internal and the external matching up.

Funnily enough, in the kitchen I'm much more externally motivated. In maths I'm very internally motivated.

Here's a small mathematical example. If I give you the following numbers:

$$25,\ 50,\ 75,\ 100,\ 3,\ 6$$

you could mess around and see what other numbers you can make, by adding subtracting, multiplying and dividing, like on *Countdown*. That would be like an *internal* motivation, where you start with some ingredients and see what you can build with them.

Or if you were actually on *Countdown*, you might try to use these numbers to make a given number, such as 952, like the mathematician James Martin did rather spectacularly some years ago, as follows:

$$\frac{(100+6)\times 3\times 75-50}{25}=952.$$

That was like *external* motivation, where you try to build something specific in whatever way you can.

Tourism

Using a map vs following your nose

When you're visiting a new city, do you set out to look for particular attractions that you've heard about, or do you just plonk yourself in the middle of the city and follow your nose? People often say that their favourite thing about a holiday was when they were just wandering around and discovered some little hidden gem down a backstreet. Sometimes this happens when you're trying to get to the Eiffel Tower or the Empire State Building or some other much-trumpeted destination, and you stumble upon a fantastic little cafe on the way.

Maths is like this too. A lot of maths happens by trying to answer a particular question or solve a particular problem. That is, you have a particular destination in mind and you just want to get there. This is external motivation. Many of the great problems in the history of maths have been like this: a particular question that needs answering, and nobody really minds how it's answered as long as it gets answered.

One of the problems with learning maths at school is that almost everything – or maybe everything – is *externally* motivated. You're always just trying to solve a problem, and worse, it's a problem that somebody else set for you, which you probably have no need to solve apart from for your maths homework, or maths exam, or something.

Take solving quadratic equations. You might remember from your past, or from Chapter 2, that if you're given an equation like

$$ax^2 + bx + c = 0$$

the solutions are given by the formula

$$x = \frac{-b \pm \sqrt{b^2 - 4ac}}{2a}.$$

This formula was produced *just* for solving that equation. It's not exactly something that you'd just come up with for fun and think 'I wonder what I can do with this?'

In real research maths, it often happens the other way, where you just give yourself a starting point in the mathematical world, and see where it takes you. I call this 'internal motivation'. It's a bit less dramatic, and so tends to get less attention. Just as your little gem down a backstreet is much less dramatic than the Eiffel Tower, and probably won't get a mention in the guidebooks. But what is it that makes Paris what it is – the Eiffel Tower or all the little gems down backstreets? Surely both, and indeed, the way they are juxtaposed.

One of the most famous instances of this is that the study of prime numbers was not thought to have any useful applications for hundreds of years. And yet, mathematicians were fascinated by them just because they're intrinsically fascinating, and seem so fundamental. How could they have known that a theorem proposed by Fermat in 1640 and proved by Euler in 1736 would become the basis for internet cryptography several centuries later? Even computers were hundreds of years away. Incidentally this is the same Fermat of 'Fermat's last theorem' fame, but the theorem in question is known as 'Fermat's little theorem' to distinguish it from the 'big' one.

In fact, Fermat's last theorem itself is an example of the curious ways in which the internal and external motivation can interact. First, there are the discoveries you can make along the way to the question you're trying to answer. Along the way to proving Fermat's last theorem, Andrew Wiles made many important discoveries about elliptic curves – a particular type of curve that doesn't sound like it should have anything to do with Fermat's last theorem. Remember, this theorem says it is impossible to make the equation

$$a^n + b^n = c^n$$

work for any whole numbers a, b, c if n is a whole number bigger than 2.

But there's also the interaction the other way, the way that I find the most satisfying and beautiful. This is where you put yourself in the middle of a city and have in mind that you'd like to see the Notre Dame, let's say, but instead of just going straight there following a map, you follow your nose down the interesting winding streets in the way that interests you. And then, lo and behold, you find yourself at the Notre Dame. In the case of Fermat's last theorem, mathematicians were also working on elliptic curves for their own sake, in a way that also happened to help with proving the theorem.

When maths is done purely by external motivation, it might be like taking such a determined route to the Notre Dame that you end up walking up a horrible main road for ages. You could say that this is maths that is overly utilitarian or pragmatic. When it's done purely by internal motivation, you might go on a very pretty journey but never arrive at anything notable. You could say this is maths that is overly idealistic or aesthetic. When the two coincide you get a journey that is interesting in its own right, with a destination that is also interesting in its own right – the best of both worlds, and the most beautiful of mathematics.

Different areas of maths have a different emphasis. Number theory has many famous unsolved problems that mathematicians are trying to solve in whatever way they can. Category theory is a bit different. One of its aims is to find the internal motivation behind everything, or to find the point of view that illuminates the internal motivation that was secretly already there. In Part II we'll see various ways in which category theory does this. Here's an example. We can think about all the possible factors of 30, that is, all the whole numbers that go into 30 without leaving a remainder. These are:

$$1,\ 2,\ 3,\ 5,\ 6,\ 10,\ 15,\ 30.$$

However, just listing them all in a row like this is not as illuminating as it might be, because in fact some of these factors are also factors of each other. If we draw lines between all the ones that are factors of each other we get a picture like this:

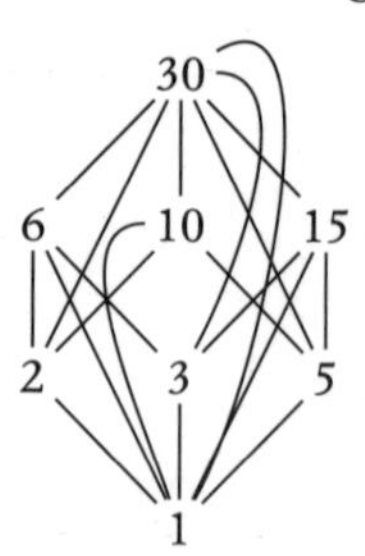

But this is a bit of a mess. We can clear it up if we decide only to draw lines where there isn't another factor *in between*. So we'll put a line between 6 and 30, but not a line directly from 2

to 30, because 6 is in between. In that case we get this more satisfying picture:

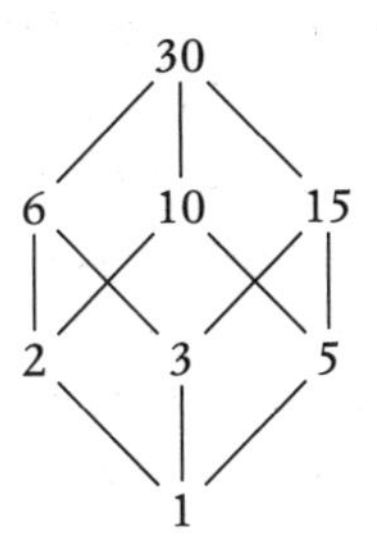

We'll come back to this sort of picture later and see that this is exactly how category theory brings out structure, making concepts visible in geometrical diagrams.

Jungle

Invention vs discovery

Sometimes I think about how different the world of 'research' was when there were still parts of the earth unmapped, still new large animals to be discovered – at least by Europeans. I suppose there are still new insects and bacteria and plants being discovered, but imagine being the first Europeans to see a platypus. And nobody believed them – when the specimen and drawing arrived in Great Britain, in 1798, it was suspected of being a hoax, perhaps created by a skilled taxidermist attaching a duck's beak to some other animal.

Here's some maths that some people think is a hoax. People often say to me 'Maths is always just right or wrong, I mean $2 + 2$ just *is* 4.' And yet, I'm now going to explain that sometimes $2 + 2 = 1$.

Do you think I'm just pulling your leg? I'm actually not. There is a world of numbers in which this is true. It's like being on a three-hour clock instead of a twelve-hour clock. We're quite used to the fact that if it's now 11 o'clock, then two hours later it will be 1 o'clock. In other words,

$$11 + 2 = 1.$$

If we were on a three-hour clock

then two hours later than 2 o'clock it would be 1 o'clock. In other words

$$2 + 2 = 1.$$

This example might seem a bit contrived, like I'd invented it for the sole purpose of making a silly answer for 'two add two'. That is, I made it with *external* motivation. But later on we'll see that this 'three-hour clock' number system arises quite naturally from *internal* motivations, and is quite important.

Here's an internally motivated example of a strange mathematical creature. You probably remember what the graph of $y = \sin x$ looks like:

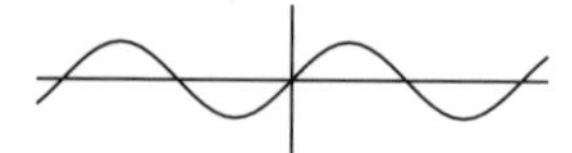

and what the graph of $y = \frac{1}{x}$ looks like:

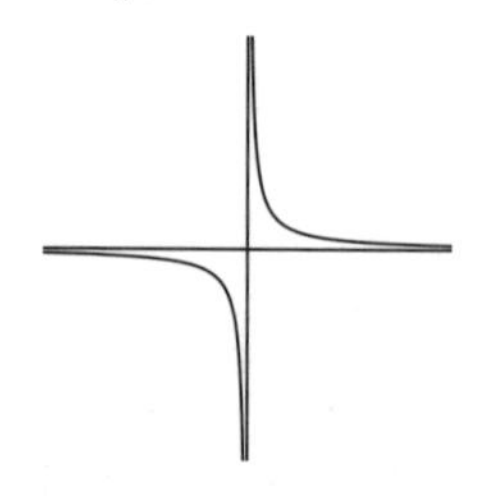

Now we might blithely try combining these, to look at the graph of $y = \sin(\frac{1}{x})$. This function is very wild.

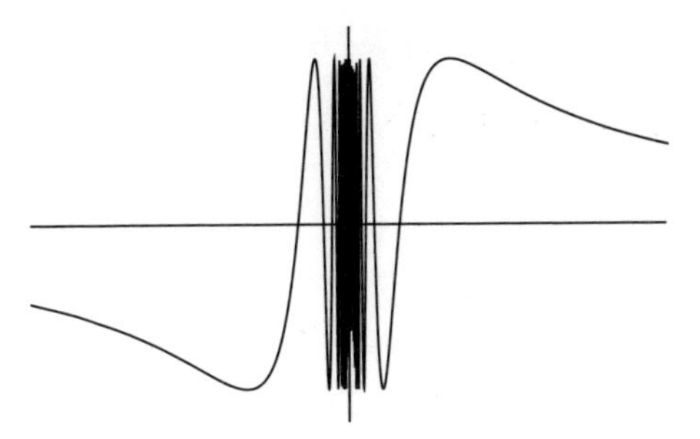

On the other hand, sometimes mathematicians set out deliberately looking for wild functions, like looking for the Loch Ness monster. What usually happens is that they want a particularly wild example of a function or a space or something, so they deliberately make one up.

Here's an example of a wild function that's been 'made up' with external motivation. We say $f(x) = 1$ if x is rational, and $f(x) = 0$ if x is irrational. This function is basically impossible to draw because it leaps up and down between 0 and 1 all the time.

An example of a space that's been deliberately made up to confuse everyone is known as the 'Hawaiian earring'. You start with a circle of radius 1, then you draw a circle of radius $\frac{1}{2}$ stuck to it somewhere on the inside:

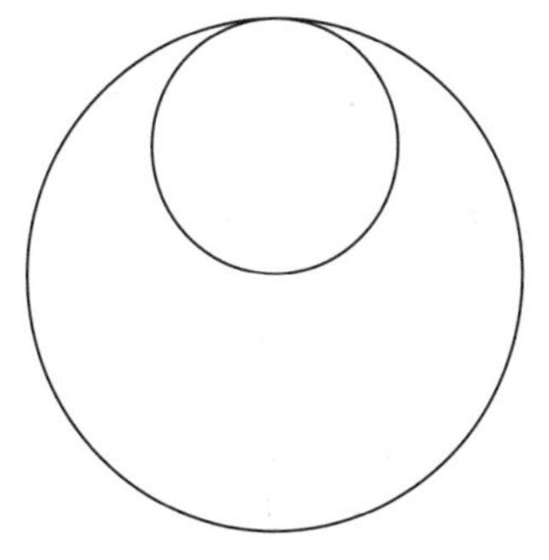

Then you add a circle of radius $\frac{1}{3}$ attached at the same point, and then a circle of radius $\frac{1}{4}$, and then $\frac{1}{5}$, and you keep going 'forever':

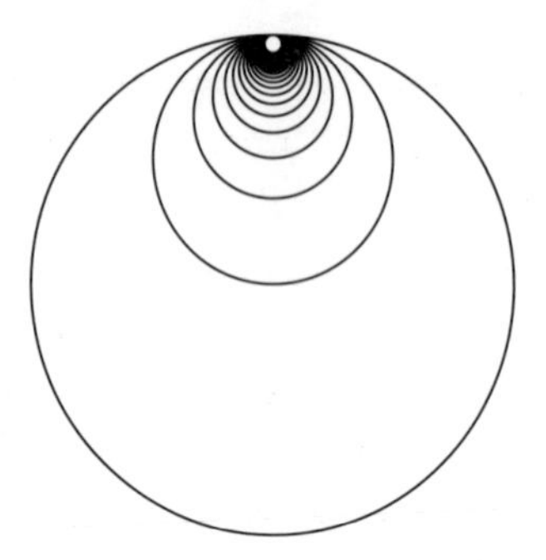

Remember, this is maths, so you don't actually have to sit there drawing forever: you just have to imagine that you did. Anyway the Hawaiian earring has very strange and wild properties which are quite exciting to topologists.

Jigsaw puzzle

Fitting pieces together vs looking at the picture

When you sit down to do a jigsaw puzzle, do you look at the picture on the box first, and match up all the pieces to the picture? Or do you put the picture away and just work out how the pieces fit together by comparing them to each other?

If you use the picture on the box, that's like an external motivation in maths. You have a clear aim, you know what the aim is, and you're trying to get there. If you don't look at the picture, that's like internal motivation. You're trying to see how the pieces fit together based on their own structure and their relationships with each other, not their relationships with something external.

I've found that a small child's first instinct is often the internal rather than the external, with jigsaw puzzles. They seem more likely to just keep trying to fit pieces with each other if they look vaguely similar, rather than comparing the pieces with the picture on the box. In fact I've found it quite hard to persuade small children there's any point at all in looking at the picture on the box; I suppose there is some stage of development where they make the connection with the internal and the external. There's also a more literal sense in which they seem more interested by the internal than the external: they tend to start with the middle of the puzzle, where the interesting part is. Most adults learn, at some point when they're growing up, that the sensible way to start a puzzle (at least, assuming it's rectangular) is to find the four corners, and then find all the edge pieces, and put the edge in place. Children, at least the children I know, don't seem to want to do that at all.

When I took physics A-level we were given a formula sheet that made the whole thing more like a jigsaw puzzle than a test of physics knowledge. So we had a list of helpful formulae that we weren't expected to remember, such as:

force between two point charges	$F = \frac{1}{4\pi\varepsilon_0} \frac{Q_1 Q_2}{r^2}$
force on a charge	$F = EQ$
field strength for a uniform field	$E = \frac{V}{d}$
field strength for a radial field	$E = \frac{Q}{4\pi\varepsilon_0 r^2}$

Now, I'll be the first to admit that a lot of it didn't really mean anything to me. In fact I was quite proud of the fact that I found a way of doing extremely well at physics A-level without really having to understand any physics. I just read the question, wrote down all the letters corresponding to the quantities given in the question, and then scanned the formula sheet for a formula containing all the correct letters. This is like the efficient adult way of doing a jigsaw puzzle by 'external' processes rather than 'internal' ones. I felt I had worked out the most efficient way to get an A at physics A-level with the least possible work.

Later we'll see that category theory often bridges the gap between internal and external processes. It makes the internal processes more geometrical, so that sometimes it really is like fitting a jigsaw puzzle together.

Here's an example of a jigsaw puzzle in category theory. You can try fitting the pieces together even without knowing what they mean. We have these two pieces:

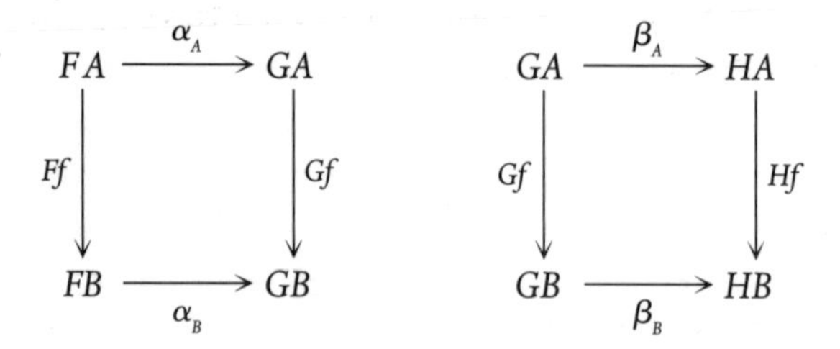

and we want to make this picture:

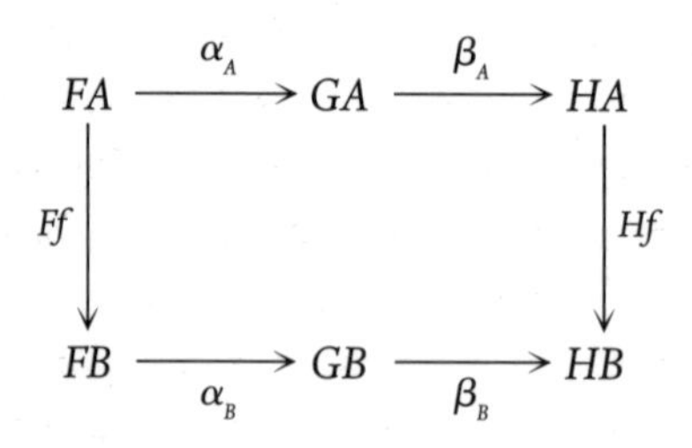

We can just fit the two pieces together sideways to make the picture, like this:

$$\begin{array}{ccccc} FA & \xrightarrow{\alpha_A} & GA & \xrightarrow{\beta_A} & HA \\ {\scriptstyle Ff}\downarrow & & \downarrow{\scriptstyle Gf} & & \downarrow{\scriptstyle Hf} \\ FB & \xrightarrow[\alpha_B]{} & GB & \xrightarrow[\beta_B]{} & HB \end{array}$$

This is a very typical calculation in category theory. The pictures get bigger and bigger and there are more and more pieces we get to use. However, because the pieces are *abstract* pieces, we have an endless supply of them and we can use each one as many times as we want.

In case you're wondering, this is part of the proof that *composing natural transformations component-wise yields another natural transformation*. More generally, this sort of jigsaw puzzle in category theory is called 'making diagrams commute', and is something I find fun and satisfying.

Marathon

Getting fit vs training for a race

If you work out or do something to keep physically fit, are you always training for a specific event? Some people always aim for a specific event, like a marathon, a triathlon or an expedition, to keep themselves motivated. Others do it for general fitness, enjoyment or stress release. Of course, it's probably some kind of combination of those things – if you don't enjoy it in the first place, then aiming for a marathon is hardly going to help.

When I ran the New York Marathon I had to change my workout dramatically. I had read various articles saying that you can run a half marathon without really specifically training for it, but not a marathon. Indeed I had already run the London Half Marathon without specifically training, other than my usual every-other-day gym routine. Also, I had a reasonably fit friend who tried to run the New York Marathon without specifically training, and he had damaged his knee.

So I did much longer workouts, building up my stamina, and following a pattern of fortnightly long runs that I found online somewhere, tapering off in the last few weeks so that the longest long run occurred something like a month before the actual marathon. It all worked fine, and I finished in exactly the time I planned (which was in fact extremely slow, but I had very realistic expectations of myself).

This is all to say that for about six months my workout became externally motivated – I had a specific aim in mind and everything was geared towards that aim. Before that, however, and ever since, my workout has been internally motivated, without a specific aim ('general fitness and weight loss' not counting as a specific aim here). The point was the workout itself, and how much I enjoy that process in its own right.

Maths is often sold for its external motivations – it is useful for getting a job, it is useful for real-life situations. But just as with the marathon, if you don't enjoy it in the first place, then imposing some contrived 'real-life' situation on it won't help. Take this example that a friend of mine gave me recently – she was trying to help her son with his homework, but needed help herself.

> *George drove 764 miles last week and his car used 15 gallons of petrol. If George averages 54 miles per gallon on motorways and 31 miles per gallon in town, how many miles did he drive in town?*

The sad thing about this question is that it *tries* to give an external motivation, but the scenario is completely contrived. Why would you need to know how many miles George drove in town unless you're his wife and trying to see if he's having an affair? And otherwise wouldn't it be easier to remember how far you drove on motorways and just subtract that from 764?

However, the internal motivation behind this question is much more interesting to me. This problem has two unknown quantities: the number of miles driven in town and the number of miles driven on motorways. It also has two pieces of information relating them: the total miles driven and the total petrol used. This is a jigsaw puzzle that has the right number of pieces.

The first step is abstraction – turning the wordy problem into a piece of maths with some letters, numbers, equations, and so on. If we write M for the number of miles driven on motorways and T for the number of miles driven in town, we can then turn our two pieces of information into equations.

* The total miles driven is 764, which means

$$M + T = 764.$$

* On motorways he gets 54 miles to the gallon, so the number of gallons used on motorways is $\frac{M}{54}$.
* In town he gets only 31 miles to the gallon, so the number of gallons used in town is $\frac{T}{31}$.
* The total gallons used is 15, which means if we add up the gallons used on motorways and in town, we should get 15, that is:

$$\frac{M}{54} + \frac{T}{31} = 15.$$

So we have two unknowns and two equations governing them. Intuitively you probably realise that if we had 200 unknown quantities and only one equation governing them, we would have not nearly enough information to work out what all the unknown quantities are. But in general if we have the same number of equations as unknowns, then we're in good shape.[1]

Personally I think that actually finding the answer at this point is the least interesting part, but that's because I particularly enjoy the process of abstraction, and enjoy that more than the process of doing calculations. In fact, did you recognise this situation from earlier in the book? Now that we have turned George's situation into two linear equations, it's just another example of the pair of equations we looked at in Chapter 2, which came from a question about my father's age. We have

[1] There are two potential problems though: the two equations could be contradictory, or they could be essentially the same. We won't go into that here.

abstracted far enough to get to a situation that we've already solved, so we definitely don't need to do any more work.

But here's the calculation in any case.

Start with the second equation: we get rid of the fractions by multiplying by 54 and 31, to give

$$\begin{aligned} 31M + 54T &= 15 \times 54 \times 31 \\ &= 25\,110. \end{aligned}$$

Now subtracting T from both sides of the first equation we get

$$M = 764 - T$$

which we can now substitute in to get

$$\begin{aligned} 31(764 - T) + 54T &= 25\,110 \\ \text{so } 23\,684 - 31T + 54T &= 25\,110 && \text{(multiply out the bracket)} \\ 23\,684 + 23T &= 25\,110 && \text{(gather the } T\text{'s)} \\ 23T &= 25\,110 - 23\,684 && (-23\,684 \text{ from both sides}) \\ &= 1426 \\ T &= 1426 \div 23 && \text{(divide both sides by 23)} \\ &= 62. \end{aligned}$$

So the answer is that George drove 62 miles in town. Bully for him. Perhaps he was having an affair?

Dreaming up some new mathematics

All through this chapter I've been discussing two different ways of coming up with a new piece of mathematics. There's the internal way, where you follow your nose, dig inside your imagination and dream up something that feels good or makes sense. And then there's the external way, where you have a specific problem that you want to solve, and so you build the tools to solve it.

We'll now compare these two approaches to come up with the notion of *imaginary numbers*.

The internal way

You might remember being told the important rule that you 'can't take the square root of a negative number'. The reason is that a positive number times a positive number is positive, but a negative number times a negative number is also positive. So if you times a number by itself it is always positive (or zero). That means whenever you *square* a number, the answer will never be negative. Taking a square root is the reverse of the process of taking a square. So to find the square root of a negative number, we have to find a number whose square is negative – and we've just decided there aren't any.

The key to the internal motivation at this point is to feel a bit dissatisfied, frustrated, irritated or even outraged that you can't take the square root of a negative number. Imagine seeing a sign saying you're not allowed to do something that you think is completely harmless – do you immediately want to do that thing? Similarly, you're now faced with a sign saying you're not allowed to take the square root of a negative number. But what harm would it do? In mathematics, 'harm' means 'causing a logical contradiction'. If something doesn't cause a logical contradiction, you might as well do it.

Now, the only way that taking the square root of a negative number would cause this kind of 'harm' would be if you tried to claim the answer was a positive or negative number, because we know that this cannot be true.

So how can there possibly be a square root of, say, -1? Well, what if there was a whole different type of number such that when you times it by itself the answer is a negative number. You might immediately say – but this doesn't exist. Just like the platypus?

The key in maths is that things exist as soon as you imagine them, as long as they don't cause a contradiction. Having a square root of -1 is not a contradiction, as long as it's a completely new number, and not any of the positive or negative numbers we already knew about. It's like having a completely new Lego piece. To make sure we don't get it mixed up with our old numbers, we call it something completely different: i. This letter i stands for 'imaginary', because it's some kind of new number that isn't 'real'. We'll come back to it later.

The external way

A more 'external' way to come up with imaginary numbers is by trying to solve quadratic equations. Remember a quadratic equation is one involving x and x^2, like

$$x^2 + x - 2 = 0$$

or

$$2x^2 - 7x + 3 = 0.$$

You might be able to remember how to go about solving these, that is, finding all the values of x that make the left-hand side equal 0. Or if you don't remember, I can tell you the answers and you can just check that substituting $x = 1$ or $x = -2$ makes the first equation true, and for the second one $x = 3$ or $x = \frac{1}{2}$ will do. Moreover, if you try any other number, it won't work.

But what about this one?

$$x^2 + x + 1 = 0.$$

No matter what number you put in, positive or negative or 0, you are doomed – the left-hand side can never equal 0. At this point you might shrug and say you never really cared about solving quadratic equations anyway. But mathematicians don't like leaving problems unsolved. Coming up with 'imaginary numbers' is a way of fabricating solutions to the equations that previously had no solutions. In this case the internal and the external have got quite close to meeting up.

Do you think it's cheating to solve a problem by inventing a whole new concept and declaring it to be the answer? For me this is one of the most exciting aspects of maths. As long as your new idea doesn't cause a contradiction, you are free to invent it. The key is to balance out the external and internal motivations for it. If you invent a new concept that is obviously contrived only to solve one problem, then it's unlikely to be a good mathematical concept in the long run, even though it won't actually be *wrong*. The best new mathematical inventions are the ones that make internal sense and also solve some existing problems.

7 AXIOMATISATION

Jaffa cakes

Ingredients

Small round flat plain cakes

Marmalade

Melted chocolate

Method

1. Put a little dollop of marmalade on each cake.
2. Use a small spoon to spread a thin layer of chocolate over the marmalade and cake.
3. Set in the fridge.

This recipe could be construed as being rather unhelpful – what kind of ingredient is 'small round flat plain cake'? What if you want to make Jaffa cakes from scratch? Then the ingredients would be: eggs, sugar, flour, butter (for the cake), oranges and sugar (for the marmalade), cocoa butter, cocoa powder and sugar (for the chocolate). Or does chocolate count as a basic ingredient?

The question of what counts as a basic ingredient and what needs to be made from more basic ingredients is a bit subtle. It depends on what you're trying to achieve. Maybe for you the Jaffa cakes themselves are the basic ingredient, and you would just buy them from the supermarket in a packet. However, I find making things myself very satisfying, and love making my own Jaffa cakes from eggs, sugar, flour, butter, oranges and chocolate.

One of aims of mathematics is to do things 'from scratch'. A consequence of asking 'Why? Why? Why?' repeatedly is that you have to boil things down to more and more basic concepts. There is always the question of what counts as a basic ingredient and what needs to be broken down further. As I have mentioned before, in maths the basic ingredients are called axioms and the process of breaking something down into its basic ingredients is called axiomatisation.

In the end mathematics is simply about things that are true. We ask why they are true, and we answer this question by boiling down a complicated truth into simpler ones. So at root, axioms are the basic truths that we're going to accept *in this particular situation*. It doesn't mean that they are absolute truths, or always true, or that they can never be broken down further. It just means that in this particular piece of maths we're going to use these as basic ingredients and see what happens.

Ginger cake

Do you have the ingredients in your kitchen?

Often, when I want to try a new recipe, I'll have to go out and buy some new ingredient that I don't keep in my kitchen all the time. As time goes on, this becomes less and less of a problem as I stock more and more things in my kitchen, especially for baking. But the first time I used dark muscovado sugar, for example, was in a ginger cake, and I had to go out and buy some. And then, of course, the recipe didn't use the exact amount in the packet, so I had some left over and started looking around for ways of using it up. Different people have different basic ingredients in their kitchen, and dark muscovado sugar is now something that I *do* always have in my kitchen, along with chocolate, butter and about eight types of flour. I only buy milk and eggs for specific recipes, whereas you might consider those kitchen staples instead of my strange attachment to almond flour.

As I mentioned in the internal vs external discussion, maybe you get ingredients specifically with a recipe in mind, or maybe you wander into your kitchen and start making things up (which these days gets called 'bakesperimenting'). Anyway, perhaps I'm being too mathematical here, but sometimes I wish recipe books would arrange themselves according to 'what other recipes you can make with the same ingredients, once you've gone to the trouble of buying these ones' (I suspect this exists on the internet now). Or even more subtly: what other recipes can you make with the same ingredients, and the new techniques you've just learnt?

Earlier on we introduced a 'new ingredient' which was the *imaginary number i*. We declared that this would be an entirely new number and would be the square root of −1. So all we know about this number so far is

$$i^2 = -1.$$

Your first objection is probably: 'But there is no such number!' However, what's more true is that there *was* no such number, but we've now invented one. Just like when we only have rational numbers there is no square root of 2, but then we invent one.

Now, what if we assume that this strange new number behaves a bit like other numbers in every other respect? This is a bit like in books or movies with time travel, where you try to make a story where everything is the same about human beings except that they can now travel in time.

We could try doing things like

$$\begin{aligned} 2i \times 2i &= 4i^2 \\ &= 4 \times (-1) \\ &= -4. \end{aligned}$$

So now −4 has a square root as well. In fact, now *every* negative number has a square root, because if a is a positive number with square root $\sqrt{a}$, then $-a$ has square root $\sqrt{a}\,i$ because

$$\sqrt{a}\,i \times \sqrt{a}\,i = a \times i^2$$
$$= a \times (-1)$$
$$= -a.$$

In order to understand what else happens when we invent this number i, we need to be very sure about what rules we want it to obey, that is, the *axioms* we're going to use.

Lego

Using the same bricks to build different things

When you sit down with a pile of Lego bricks, you have two things:

1. a pile of objects
2. some ways of sticking them together.

The great genius of Lego (or perhaps I should say, one aspect of its great genius) is that it is so simple and yet has so many possibilities. Analysing this genius a bit further, I think it's crucial that the ways of sticking bricks together are very clear, and there aren't too many of them.

Maths works like Lego. You start with some basic building blocks, and some ways of sticking them together. And then you see what you can build. But there are two ways you can do this:

1. you can start with the bricks and see what you can build, or
2. you can start with something you want to build, and see what bricks you'll need in order to build it.

For example, to build a Lego car, you'll probably need some wheel pieces. Unless you're building a really big one, in which case you can make your wheel pieces from basic bricks, like they do at Legoland.

This is related to the internal vs external discussion; in a way axiomatisation is an externally motivated way of dealing with an entire mathematical structure or world. It's a way of working out how to build the structure you want, using logic.

Let's try it with numbers. To make all the natural numbers, 1, 2, 3, 4, 5, and so on, you only have to give yourself the number 1 as a brick, and 'addition' as a way of sticking things together. It might take you a long time to make a million in this manner, but in maths we care first about whether you can do something *in principle*. How long it would take is a whole separate question. And after all, some millionaires made their millions one pound (or dollar) at a time, by selling very small items such as oven chips. I think this is why toddlers get so excited about learning to climb stairs, because they realise that all they have to do is learn to climb up one step, and then if they do that repeatedly they can go higher and higher and higher, perhaps all the way to the sky. (Except, usually, some killjoy adult comes along and removes them from the stairs.)

Things get quite exciting when you do Lego step 2 and then Lego step 1. That is, first you decide you want to make a car, and you get hold of all the pieces you need for that – wheels, doors, and so on. And then you see what else you can make with the same pieces – a pick-up truck, perhaps, or a space rocket.

You also might start to think about more exciting ways of sticking your bricks together. When small children start playing with Lego, you might see them just stacking bricks directly on top of each other in a big tower. It might take them a bit longer to move to stacking them offset, so they can make a wall. And then what about going round corners, so that you can build an entire house? Likewise, with numbers, once you've got bored of just adding them up, you move on to subtracting them, multiplying them and dividing them, and just like that you've invented fractions.

Axioms in maths are like the basic Lego bricks and the ways you allow yourself to stick them together. One of the ways that mathematicians set up their worlds to behave according to strict logic, is to 'axiomatise' them. That is, you decide which bricks and which ways of sticking them together you're going to allow. This doesn't mean you'll never allow yourself other bricks and methods, but just for now, you'll only allow yourself these, to explore how much you can build in this way.

The important thing is that the bricks are considered to be *basic* – you're given them in a box. You don't try and break them up, although I'm sure there are children whose first reaction to Lego is to try and smash it to pieces.

Here are some axioms for the integers.

* You can add any two integers and get another.
* If a, b and c are any integers, then $(a+b)+c=a+(b+c)$.
* If a is any integer, then $0+a=a$.
* For each integer a, there is another one b such that $a+b=0$.

The last rule means we know we must be talking about the integers and not just the natural numbers, because it's really telling us about negative numbers. But we could also be talking about the 'three-hour clock'. You might think we don't appear to have negative numbers, because we only have the numbers 1, 2 and 3, on this clock. But each of these numbers does have a partner that makes it add up to 0 on the clock, once we remember that 0 is the same as 3:

$$1+2=3$$
$$2+1=3$$
$$3+3=3.$$

These axioms are actually the axioms for the mathematical notion of a *group* and we will see that there are plenty of other examples of groups, including things that have nothing to do with numbers.

Doctors-and-nurses football

Imposing careful rules so that strange loopholes don't arise

A doctor friend of mine once told me about a doctors-and-nurses football tournament they were having at Addenbrooke's Hospital in Cambridge. Apparently it was with mixed teams, and teams started with a bonus of one extra goal per female member of their team. It turned out that one team realised they had more women than anyone else and just stood the entire team in goal for the whole match.

Do you think that decent people should keep to the spirit and not just the letter of the rules? Or do you think that rules should be sufficiently watertight not to let such peculiar loopholes occur?

In maths, we are dealing with objects that only obey the rules of logic. So we cannot possibly ask them to interpret the spirit of the rules rather than the letter of the rules. The 'letter' of the rules is what happens if you follow them by exact logic, and so it is the only thing our mathematical objects will do. So when we make those rules, we have to be careful to close the loopholes ourselves.

Here's an example of a mathematical loophole that can be confusing. Remember that a *prime number* is one that is 'only divisible by 1 and itself'. However, we have to add a caveat, almost like an afterthought, and declare that the number 1 doesn't count as prime.

Sometimes this gets explained by something like 'Well a prime number is one that has exactly two factors, whereas 1 only has one'. This is true, but doesn't explain *why* we need this rule. The key is to understand what prime numbers are there for – they are the building blocks of numbers, where we are trying to build numbers by multiplication rather than by addition. If we're building by addition, we only need the number 1 and we can keep adding it up to get all the other numbers. If we're building by multiplication, the number 1 doesn't get us anywhere, because if you multiply things by 1 nothing happens. This means it's not a very good building block.

More technically, we want to be able to say that every whole number is a product of prime numbers in a *unique* way. For example, there is only one way of building the number 6 by multiplying prime numbers, which is 2×3 (the order doesn't matter, so 3×2 counts as the same thing). However, if we said 1 counted as a prime number, we'd be able to do $1 \times 2 \times 3$ and $1 \times 1 \times 2 \times 3$, and so on. The 1 would ruin everything, without helping us in any way at all. So we have to rule out this loophole.

Democracy

Imposing careful rules can have strange effects

There is no such thing as a fair voting system.

You might have a hunch about this, or you might believe it vehemently, based on your experience of elections. But it's also a mathematical theorem.

The thing is, to make sense of this statement we first have to be precise about exactly what we mean by fair. That is, we have to set up our axioms precisely. In this case, it's called Arrow's theorem. It's relevant not just to political elections, but also to things like competitions with a panel of judges who need to decide on a ranking of competitors.

The axioms for a fair voting system in this setting are:

1. **Non-dictatorship:** The outcome is decided by more than one person.
2. **Unanimity:** If everyone votes that X is better than Y, then X will be ranked higher than Y in the final outcome.
3. **Independence of irrelevant alternatives:** The relative ranking of X and Y should not be affected by someone changing their mind about Z.

Arrow's theorem then says that if there are more than two people (or things) to vote for, there is *no fair voting system*.

> The axiom most commonly violated by modern democratic voting systems is the third one, which is why tactical voting becomes a possibility.

You might have had the experience of having an argument with mathematical types, where the argument all ends up boiling down to definitions. For example, if you try and have an argument about whether you have a soul or not, it all hinges on what you take 'soul' to mean.

One of the aims of mathematics is to study everything using logic, and mathematicians don't want their answers to boil down to arguments about definitions. So they are careful to say exactly what they are using as their definitions in the first place, like laying down the ground rules. You might be cross when someone is disqualified from a 100 metre sprint because of a false start, but those are precise rules of the event. You might disagree with the rules, but you can't (rationally) disagree with the fact that the rules were applied.

This is one of the things that makes mathematics precise, but also one of the things that can frustrate people about it. It is very unyielding. You can think the rules are stupid but you can't do anything about them. I always thought it was annoying that squash rackets had such small heads – but that's part of the game. It's part of the axioms. Do you think it's stupid that there is an 'imaginary' number that is the square root of -1? Tough – it doesn't matter that you think it's stupid. We can play a game involving that number as a building block, and it makes no difference whether you believe in it or not – that is the game.

High jump

Imposing careful rules to remove human judgement

There's something I find very satisfying about the high jump, as a sport. Not to take part in it, mind you (as I lamented in the chapter about abstraction), but to watch. Because the rules and the aims are so clearly defined. You have to get over a bar and that's pretty much it. Now, for all I know, there are some minuscule technicalities that I'm missing here, but from the spectator's point of view that does seem to be what's going on. This is unlike, say, synchronised swimming, or wrestling, where however much effort has been put into making it as objective as possible, it still appears to come down to human judgement in the end.

Maths is about removing the human judgement from things, so that everything proceeds just by logic. This can be both satisfying, because it makes everything so unambiguous, and unsatisfying, because we are essentially removing ourselves from everything. However, the aim isn't to turn all of human experience into this process, any more than we're claiming that the high jump is the whole of life (even if it might seem like that for the competitors, while they're doing it). The aim is to study certain aspects of a situation unambiguously. With the high jump, the aim is to see how high a bar a human being can jump over with a certain amount of run-up. This is beautiful to watch (there's something so elegant about the Fosbury flop, not betrayed by its name), but also it fascinates me because it highlights one pure feature about human beings. The 100 metre sprint fascinates me for the same reason. It's not because I'm glad Usain Bolt will be able to catch a bus better than the rest of us.

You can almost imagine how the high jump was first 'axiomatised', that is, how the rules came about. Again, let's allow ourselves some historic licence here. Perhaps some people were challenging each other about how high they could jump over a fence. Perhaps one person realised that if they ran up to the fence they'd be able to jump higher. And then there was an argument about how long a run-up would be allowed. And then there was an argument about whether you're allowed to put a mattress on the other side, to break your fall. And so on.

Axiomatising parts of maths happens in a similar sort of way.

The *rational numbers* are formed from the integers by taking any fractions $\frac{a}{b}$, where a and b are integers (positive or negative whole numbers). Pretty soon you realise you need to add a clause in there saying that b isn't allowed to be 0, because that wouldn't make sense.

But then you realise you need another clause to explain that $\frac{1}{2}$ is actually the same as $\frac{2}{4}$, $\frac{3}{6}$, and so on. There are two ways you can do this. You can either declare that all your fractions have to be in their *lowest terms*, that is, the top and bottom have no common factors that can be cancelled out. However, this is a bit disingenuous, because $\frac{2}{4}$ is a perfectly good fraction.

The more mathematically mature way to do it is to say you're going to take all the fractions $\frac{a}{b}$ but impose an axiom on them to govern when they are actually the same fraction, which goes like this:

$$\frac{a}{b} = \frac{c}{d} \quad \text{whenever} \quad a \times d = c \times b.$$

This looks a bit obscure, but comes down to saying the same as 'if we cancelled both to their lowest terms, they'd be the same'. It's just a much more efficient way of saying it.

Cake cutting

Imposing careful rules to remove ambiguity

If you have a brother or sister, I'm sure you encountered this problem when you were little: how can you share the last piece of cake fairly between you? You probably hit upon the brilliant solution 'I cut, you choose!' Now, if you're the one cutting, it's up to you to cut fairly, because if you make one piece bigger than the other, your sister will obviously take the bigger one, and you'll only have yourself to blame.

That's all very well, but what if you have a brother *and* a sister, so you have to share the cake between three. Or four? Or eleven?

It's not so hard if you're sharing a round cake (you could always get out a protractor), but what if it's just one piece of cake? Or a dinosaur cake? How can you share it fairly?

The key here is just like in the question of a fair voting system: what does 'fair' mean? In order to try and solve this problem, we have to be clear exactly what the problem is, and this involves axiomatising the situation of cake cutting. This has actually been done and turned into a mathematical problem.

Let's suppose we're doing it for three people. Here are two notions of 'fairness':

1. Everyone *thinks* they've got at least one-third of the cake.
2. Nobody *thinks* anyone else has more cake than them.

The first we could think of as 'absolute fairness' because everyone just evaluates their own piece of cake by itself. The second we could think of as 'relative fairness' because now everyone is comparing their piece of cake to everyone else's. It also gets called 'envy free' because the important thing is to make sure nobody is envious of anyone else.

If you're only sharing cake between two people, these two types of fairness are the same. But with three people or more, it gets much more complicated. You might well think you have a third of the cake, but if you think your brother's got more than you, you think it's unfair, even though it shouldn't really be your problem.

The problem is turned into a piece of mathematics by stating these rules of fairness very precisely. We have to take into account various complicated possibilities. Not only might the cake not be round, but it might have different decorations on it, such as icing, marzipan, cherries, that different people like differently. When I was little my best friend and I could always share Christmas cake perfectly, because she didn't like the cake and I didn't like the icing and marzipan.

In fact once we have axiomatised the sharing of a cake so precisely, we see that we can easily apply it to sharing *anything*, including things that can't be cut up. The problem can be solved mathematically, and the solution is rather complicated. The interesting thing is it's much more complicated when envy gets involved – a mathematical proof that envy complicates the world.

Given any way of sharing the cake out between n people, everyone personally rates each piece of cake as a proportion of the whole. So if there are five people and you think the cake has been shared perfectly fairly, you'll give each piece a score of $\frac{1}{5}$ or 0.2. But if if you don't think it's fair, then maybe you'll give the five pieces of cake scores of

$$0.3,\ 0.25,\ 0.25,\ 0.1,\ 0.1$$

to show that one piece was the best (perhaps because it has a cherry on it) and two pieces were definitely short-changed. But someone else might rate the pieces differently (maybe because they hate cherries).

* **Absolute fairness** means that everyone gets a piece of cake that they rated at least $1/n$.
* **Relative fairness** means that if I rated my piece x and your piece y then $x \geqslant y$.

So in the example of my friend and the icing, I rated the cake with no icing as 1, and the icing with no cake as 0. Conversely, she rated the cake with no icing as 0, and the icing with no cake as 1. I thought my piece was much better than hers, and she thought hers was much better than mine, and so we were both happy, and friends for life.

Why? Why? Why? (again)

Where the careful rules of logic come from

When a small child keeps asking 'Why?' repeatedly, you might wonder if it's ever going to end. The answer is: no, it isn't.

Small children seem to be more bothered than we are about the fact that some things are inexplicable. As adults we get used to accepting things as true even though they're not explained, because they're given to us on some higher authority. Most of us these days accept that the earth is orbiting around the sun, but most of us have seen no evidence of this fact other than that some other people told us it's true, and we believe them. Why do we believe them? Because we trust that some other people have checked them out. But why do we believe those people?

We expect children to learn how to 'be reasonable', but we also expect them to believe things that they don't understand. I'm not surprised that this is confusing to them. The adults keep flipping randomly between things that are really parts of logic and which things are 'beliefs'.

One of the points of axiomatising a system is to make that distinction very clear. On the one hand we have our basic starting points, the axioms which are truths that we don't try to justify. On the other hand we have logical deductions leading us to other truths, justified from the starting point of the axioms.

The thing is that if we don't start with some assumptions, we won't get anywhere. Have you ever tried building something out of Lego, starting with no bricks? Of course not. Likewise, using sheer logic is all very well but it only enables you to deduce things from other things. If you start with nothing, you get nothing. So maths isn't about 'absolute truth' after all, as described in the following paradox by Lewis Carroll, first published in 'What the Tortoise Said to Achilles' in the 1895 issue of *Mind*.

Carroll considers the following three statements:

A *Things that are equal to the same are equal to each other.*

B *The two sides of this Triangle are things that are equal to the same.*

Z *The two sides of this Triangle are equal to each other.*

This is the kind of situation you might get into if you

measure two sides of a triangle with a ruler and discover that they are both 5 cm long. Does that mean the two sides of the triangle are the same length *as each other*? That is, does Z logically follow from A and B? It does seem to be rather obvious... but why? If a two-year-old asked you why, what would you say? It would be rather hard to explain. The reason this is called a paradox is that Z seems so obviously true, once you know A and B, and yet, logically there is no way of deducing it from *only* A and B. It only follows because we believe the following statement:

C *If A and B are true, Z must be true.*

Now does Z follow? Only because we believe this:

D *If A and B and C are true, Z must be true.*

Now does Z follow from A, B and C...? Oh dear. We seem to have got ourselves into a situation where we need an infinite number of steps to get to Z, although Z is 'obviously' a consequence of A and B. This is why it's called a paradox.

You might want to hit me now, and say that Z *just does* follow from A and B. Actually that's what mathematics does as well. It accepts as a basic principle that once you know that P is true, and if you also know that 'P implies Q', then you are allowed to conclude that Q is true.

In Carroll's paradox, P would be 'The two sides of this triangle equal the same thing', and Q would be 'The two sides of this triangle equal each other'.

In mathematical logic, this basic principle is called a *rule of inference* because it allows us to infer something from something else. It is given the grand name *modus ponens* (literally, 'method of affirming') and is so basic and obvious that it can be hard to remember that it's really an axiom, an ingredient that we're allowing ourselves to use. It's like when you don't count salt and pepper as ingredients in a recipe because they're so basic. If the 'paradox' still doesn't seem like a paradox to you, this might be evidence of just how deeply basic this rule of inference is in our logical thinking.

All of mathematics can be seen as a process of starting from some basic assumptions A, B, C, and so on, and trying to use logic to get to some final conclusion Z using the rule of inference. To help us understand how to do this correctly, we'll now look at a couple of ways it can go wrong. One is where you start with correct assumptions but the wrong process of deduction. This is like using the right ingredients but the wrong method in a recipe. But first we'll look at a case where even the basic assumptions turn out to be wrong.

Helicobacter

When you have the right rules but the wrong building blocks

The 2005 Nobel Prize in Physiology or Medicine was awarded to Barry Marshall and Robin Warren for their discovery of the bacterium *Helicobacter pylori* in the stomach, and their work on its role in gastritis and ulcers. In his Nobel speech, Warren describes the difficulties he faced in convincing the world that this bacterium really was in the stomach. He said:

> *Since the early days of medical bacteriology, over one hundred years ago, it was taught that bacteria do not grow in the stomach. When I was a student, this was taken as so obvious as to barely rate a mention. It was a 'known fact', like 'everyone knows that the earth is flat'.*

It appears that the medical community was taking this as an axiom, something that did not need justification. As sensible as assuming that the earth is flat. Warren goes on:

> *As my knowledge of medicine and then pathology increased, I found that there are often exceptions to 'known facts'.*

That is to say, sometimes the axioms turn out to be wrong. One of the points of clearly expressing your axiomatisation of a system is so that you know which facts might need to be challenged. Just like when Euclid axiomatised geometry, enabling mathematicians to think clearly about parallel lines, which in turn enabled them to come up with the different forms of geometry that we talked about earlier.

Cot death

When you have the right building blocks but the wrong rules

In 1999 a lawyer Sally Clark was wrongly convicted of the murder of her two baby sons. The conviction was largely based on the 'expert evidence' provided by the psychiatrist Roy Meadow. The question was whether the two babies had each died of Sudden Infant Death Syndrome, or whether that was too much of a coincidence. Meadow pronounced that the probability of two cot deaths occurring in the same family was 1 in 73 million. The fatal flaw was that Meadow had come to this conclusion simply by squaring the probability of one cot death occurring.

Now, in many situations, this is the correct way to calculate the probability of something happening twice. If you toss a coin, the chance of getting a head is supposed to be half. If you toss it twice, the chance of getting heads twice is

$$\frac{1}{2} \times \frac{1}{2} = \frac{1}{4}.$$

However, if you tossed it a thousand times and got heads every single time, you might begin to suspect that the coin was weighted, and that the chance of getting a head wasn't half at all. You would suspect the coin of being somewhat predisposed to landing on heads.

With certain illnesses you don't need a thousand cases in one family before suspecting that the probabilities are likewise not so straightforward. If one person in your family has the flu, you're much more likely to get it, just because it's infectious.

And if an illness has some genetic component, the same is true – for example, if one female in a family has breast cancer, the other females are much more likely to get it. This isn't because it's infectious, but because the presence of one case is already enough to indicate that the family is more prone to breast cancer.

Technically, this tells us that occurrences of breast cancer in members of the same family are not *independent* events. Probability can only be calculated by simple multiplication if events are independent.

The assumptions of Roy Meadow appeared to be this.

A The probability of a cot death is (approximately) 1 in 8500.

B The probability of two independent events occurring is obtained by squaring the probability of one event occurring.

Z Therefore the probability of two cot deaths in one family is 1 in 8500 squared, that is, 1 in 73 million.

But in fact there was a suppressed assumption:

C Cot deaths in a family are independent events.

At the time, assumptions A and B seemed irrefutable, and therefore Z was accepted. But professional statisticians immediately spotted the flaw and the Royal Statistical Society issued a press release to draw attention to it. Being illogical might be dangerous, but sometimes it is even worse to apply logic incorrectly, and give oneself an air of scientific truth that is then difficult for non-experts to refute. Sally Clark's conviction was overturned but not until 2003 when she had already spent three years in prison for double murder. She never recovered from her traumatic experiences, and died of alcohol poisoning four years later.

Chess

Simple rules, complex game

One of the enduringly fascinating things about chess is that the rules are not that difficult to explain, but the resulting game is ferociously complex. I recently explained the rules to a six-year-old and we got playing within seconds, aided by the fact that the computer version of the game helped him know where any given piece could, notionally, move to.

One very satisfying thing about making rules for a game, or axiomatising a system, is to see how few rules, or how few axioms, you can start with, and still produce a really complex game. This is like when mathematicians tried to show that the parallel postulate was redundant to Euclid's geometry, as we discussed in Chapter 5. If one of your rules can be deduced from some others, then you don't need to say it out loud.

One of the very appealing things about category theory, in mathematics, is that you don't need very many rules to get started. Just like maths, category theory can seem difficult for (at least) two reasons:

1. Perhaps you don't know about or don't care about the examples you're trying to illuminate. This is a problem if you're more externally motivated than internally motivated.
2. It uses very few assumptions, so it seems like you have to work harder to get anywhere. This is a bit like doing a jigsaw puzzle with very tiny pieces. Or making a recipe from scratch instead of using a mix.

The second point is a bit like the question of sports where very little equipment is needed (for example, running) compared with sports where very expensive equipment is needed (for example, sailing). Unsurprisingly, richer countries do rather better in sports involving expensive equipment. However, I am much more interested in sports without this equipment, both to watch and also as a study of human behaviour. Yes, it's much harder to run 10 km than to do it on a bicycle, but the fact that the competitors are only relying on their own bodies is very exciting.

Likewise, I find mathematics the most exciting of all subjects, because it only relies on brainpower.

Number systems, clocks and symmetry

Examples of axiomatisation

I'll now demonstrate an axiomatisation of number systems that enables us to bring 'clock' arithmetic and also symmetries of shapes into the same frame of reference. It is the mathematical notion of a *group*.

First of all we declare we have a set of 'objects'. At this point it doesn't matter what those objects are – what matters is that they satisfy the rules we're about to impose on them. In the end we'll see that they could be whole numbers, fractions, symmetries of a triangle, and many other things. It won't work if we only take positive numbers or irrational numbers. It won't work if we take birds or cars or apples.

Next we declare that we have a way of combining any two of our objects and producing a third object of the same type. For numbers this could mean adding them together or multiplying them. We could try doing subtraction but we'll see in a minute, when we examine the rules, that this won't obey all the rules.

This 'way of combining' objects is called a *binary operation* because we take two things and perform an operation on them to produce a third. In more abstract situations this operation might not look like combining the objects at all – it's just any process that produce a third object as the answer. We might write this as $\circ$ in general, because we don't know if it's actually going to be $+$ or $\times$ or something else entirely, but we need to write it as something when we write down what the rules are that it has to obey. Here are those rules.

Associativity

For any three objects a, b, c, the following equation must hold:

$$(a \circ b) \circ c = a \circ (b \circ c).$$

So for addition this would say things like

$$(2 + 3) + 4 = 2 + (3 + 4),$$

and for multiplication this would say things like

$$(2 \times 3) \times 4 = 2 \times (3 \times 4).$$

The 'abstract' formulation using a, b, c and the funny symbol $\circ$ has saved us a lot of effort, because not only can we avoid having to write down this equation for every single number (which would be impossible as there are infinitely many of them) but we don't even have to write this down for addition and multiplication separately, as they are both examples of the same concept.

We can now see that subtraction won't work. Because for example

$$5 - (3 - 1) = 5 - 2 = 3$$

but

$$(5 - 3) - 1 = 2 - 1 = 1$$

so associativity does not hold.

Identity

There has to be an object that 'does nothing'. We could call it E, and then this means that for any object a,

$$a \circ E = a \quad \text{and} \quad E \circ a = a.$$

The object E is called the *identity* or sometimes the neutral element.

If we're talking about numbers and addition, can you work out what the identity element would be? It has to be a number such that when you add it to anything else, nothing happens. So it has to be 0.

What about if we're doing numbers and multiplication – it has to be a number such that when you *multiply* anything else by it, nothing happens. So it has to be 1.

This is another reason that we can't do this with just the irrational numbers – because there is no irrational number that could be the identity element.

Inverses

Every object has to have an *inverse* object, so that they can cancel each other out. Technically what this means is that when you combine them, the answer has to be the *identity* element. So for every object a there has to be an object b such that

$$a \circ b = E \quad \text{and} \quad b \circ a = E.$$

Can you work out what this means if we're doing numbers and addition? Remember that the identity element here is 0, so for any given number a we need another number b such that

$$a + b = 0 \quad \text{and} \quad b + a = 0.$$

If this is too abstract for you, try it with an actual number, say 2. What number is there that we can add to 2 to make 0? The answer is -2. And this will work for any number a, as we can always add it to $-a$ to get 0. It's worth remembering at this point that this will even work for negative numbers. If we start with -2 then the number we want to add on to get 0 is 2, but this is the same as $-(-2)$.

This is the reason we can't do this with positive numbers, even if we include 0, because we won't be able to get these inverses to work.

What about if we're doing numbers and multiplication? In that case the identity element is 1, so for each number a we need another number b such that

$$a \times b = 1 \quad \text{and} \quad b \times a = 1.$$

Again, you might like to try this with the number 2 again. What number can we times by 2 to get 1? The answer is $\frac{1}{2}$. At this point we should realise two things. First of all, we can't do this with whole numbers – we need fractions. Secondly, we can't do this with 0, because it is not possible to multiply 0 by anything and get 1 as the answer, because the answer will *always* be 0.

Examples

Now that we've axiomatised the notion of a *group* we can say what some examples are. In each case we have to say what the set of objects is, and what the method of combining them is.

* The integers with *addition* is an example, but the integers with *multiplication* is not, because there won't be inverses.
* The rational numbers with addition is an example, but the rational numbers with multiplication is not, because 0 won't have an inverse.
* The irrational numbers with addition is not an example, because addition is not even a valid binary operation on irrational numbers – if you add two irrational numbers, you might get a rational number as the answer. For example, we could try adding $\sqrt{2}$ and $-\sqrt{2}$ and of course we get 0, which is rational. Do you think this is 'cheating'? It might be an annoying example, but in maths we stick to rules very pedantically, whether it's annoying or not.
* The natural numbers (positive whole numbers) is not an example with addition or with multiplication, because there won't be inverses.
* The natural numbers with subtraction is not an example, again because this is not a valid binary operation on

natural numbers. For example, 1 and 4 are natural numbers, but $1 - 4 = -3$, which is not a natural number. Subtraction is a binary operation on integers, but as we saw above it does not satisfy the associativity rule, so this operation doesn't make the integers a group.

* Three-hour clock arithmetic is an example: the set of objects is just the numbers 1, 2, 3, and the way of combining them is three-hour clock addition. We can do this with any number of hours as well, to give the n-hour clock. Mathematically this is called the 'integers modulo n'. Arithmetic on a clock face is then called *modular arithmetic* and we'll come back to it several times as it's an intriguing example.

Remember that matrices look like this:

$$\begin{pmatrix} 1 & 0 \\ 3 & 2 \end{pmatrix}.$$

This one is a 2-by-2 matrix as it has two rows and two columns. We can add up 2-by-2 matrices by adding the numbers up place by place. So

$$\begin{pmatrix} 1 & 0 \\ 3 & 2 \end{pmatrix} + \begin{pmatrix} 7 & 4 \\ 6 & 5 \end{pmatrix} = \begin{pmatrix} 8 & 4 \\ 9 & 7 \end{pmatrix}$$

because for the top left place we do $1 + 7$, for the top right we do $0 + 4$, and so on. Now we can look for an matrix that 'does nothing' when we try to add it to things. The matrix we need is

$$\begin{pmatrix} 0 & 0 \\ 0 & 0 \end{pmatrix}.$$

This is the matrix that plays the role of zero in the world of matrices. We can check all the other axioms to show that 2-by-2 matrices form a group under addition. This also works for any other size of matrix; we just can't mix the sizes up because then we won't be able to add them together.

Finally an example that has nothing to do with numbers, to show the power of this axiomatisation. Actually this example is where the idea of a group really came from, which is thinking about symmetry.

We have already mentioned the symmetries of an equilateral triangle.

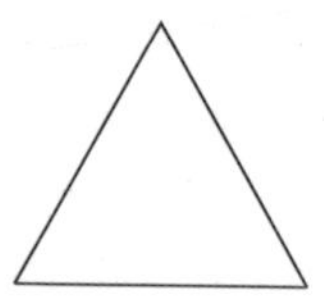

There are two kinds of symmetry: rotational symmetry and reflectional symmetry. The equilateral triangle has three of each kind.

In maths, we can think of symmetry as an *action* that you perform on the triangle. You can imagine cutting out a triangle and actually rotating it. For the reflectional symmetry you can actually flip it over along the line of symmetry. (Usually we explain reflectional symmetry as being folding something in half and the two halves matching up, but you could also imagine flipping it over and it looking the same after flipping it.)

So now we can combine these symmetries by doing first one and then another. We can imagine rotating the triangle and then flipping it. The result will have to be another symmetry. For example:

* If you rotate it and then rotate it again, you get another rotation.
* If you flip it and then flip it again, you will get back to the front but possibly a different way up, so the answer in that case is a rotation.
* If you flip it and then rotate it you will end up on the back, so in that case the answer must be a flip, that is, a reflection. Likewise if you rotate it and then flip it.

We could make a big 6×6 table showing all the possible combinations of symmetries and what the answers are if we do two in a row. Then we can check that the axioms are satisfied. The identity element is a symmetry that you probably don't think about that much: rotation through 0 degrees. If we're thinking about symmetry as an action, this means we're leaving the triangle in exactly the same place.

Now it's easier to see what the inverses would be. The inverse of a rotation is another rotation by the same amount but backwards. The inverse of a reflection is a reflection in the same line – if you do the same flip twice you get back to exactly where you started. Associativity is a bit harder to see, but if you work out all the possibilities it does turn out to work.

This means that the symmetries of the equilateral triangle form a group. In fact, the symmetries of any given object form a group. This is one of the important reasons for studying groups at all, and it shows that if you look at things abstractly you can discover unlikely similarities between them. In the end, mathematics is all about finding similarities between things, and category theory is about finding similarities between mathematical things.

8 WHAT MATHS IS

Custard

Ingredients

6 egg yolks

50 g caster sugar

1 pint of double cream, single cream or milk as desired

Method

1. Whisk the egg yolks and caster sugar until very thick, pale and creamy. If you watch carefully while whisking, you will see them change colour and get noticeably thicker, as if they've undergone a chemical change.
2. Heat the milk or cream until bubbles appear round the edge of the pan. Pour slowly into the egg mixture, stirring gently.
3. Quickly wash and dry the saucepan and pour the mixture back in. Heat on a low heat, stirring very continuously until it coats the back of the spoon.

Making custard is thought of as a tricky process. The reason is hidden in the last step of the recipe. A more accurate description of the last step would go like this.

> Watch for a thickening of the custard that looks like a qualitative change, and then take it off the heat. But don't wait until the custard is as thick as you want it, because it will continue cooking after you take it off the heat and then be overdone and probably curdle. However, if you don't wait long enough, then the custard will be thin and

> uncooked. It might help to have a glass jug ready, with a sieve over it. Apparently if you pour the custard through a sieve it will stop cooking more quickly. I've tried it both ways and am not sure if I noticed a difference, but it does make me feel reassured that I've taken every possible precaution. If you cut it very fine, then the last part of the custard in the pan will be overcooked by the time you've finished pouring, so you might want to leave the last part behind.

We can now see why custard is thought of as being difficult – the instructions are not very clear-cut. It's not like measuring ingredients, setting the oven temperature and putting on a timer. The last step requires almost an entire essay to describe it, and even then the only way to get it right is to practise it yourself. Books often say something about waiting until the custard covers the back of a wooden spoon in such a way that if you run your finger through it it leaves a mark, but I have never been able to understand this instruction, because my finger seems to leave a mark before I've even started cooking the custard mixture at all. This is one of the things I find exciting but a bit scary about making custard. You have to use your judgement, in a very short space of time, and it would be hard to get a robot to do it.

I'm now going to draw this half of the book to a close by showing that maths is *easy*, in the same sense that custard is difficult.

Logic vs illogic

Why maths is easy and life is hard

It is a truth universally acknowledged that mathematics is difficult. Or, at least, so it seems from the number of times I tell someone I'm a mathematician only for them to respond, 'Wow, you must be really clever.'

This is the Myth of Mathematics. I'm now going to take the bold step – perhaps the rash step – of exploding it. This is a bit like the Masked Magician whose TV show explained how magic tricks work – with the result that he was vilified by the Magic community. Nevertheless I am going to show that mathematics is easy, and in fact that it is precisely 'that which is easy'.

First I'd better make clear what I mean by 'easy', just like in the cake-cutting problem you first have to be clear what you mean by 'fair'. And here's what I mean: something is easy if it is attainable by logical thought processes. That is, without having to resort to imagination, guesswork, luck, gut feeling, convoluted interpretation, leaps of faith, blackmail, drugs, violence, and so on.

By contrast, life is hard. That is: it involves things that are not attainable by logical thought processes. This can be seen as either a temporarily necessary evil or an eternally beautiful truth. That is, we can think:

1. life is like that only because we haven't yet made ourselves logically powerful enough to understand it all, and that we should be continually striving for this ultimate rational goal, or
2. we will never be able to encompass everything by rationality alone, and that this is a necessary and beautiful aspect of human existence.

I'm in the second camp. Here's why.

Mathematics is easy

As long as you have the right definition of 'easy'

What is mathematics? Earlier on I said: 'Mathematics is the study of anything that obeys the rules of logic, using the rules of logic.' What is mathematics for? I'll sum up the discussion of this first half of the book as follows. Maths has two broad purposes:

1. To provide a language for making precise statements about concepts, and a system for making clear arguments about them.
2. To idealise concepts so that a diverse range of notions can be compared and studied simultaneously by focusing only on relevant features common to all of them.

Put more simply, mathematics is there to make difficult things easier. There are many reasons that 'things' can be difficult and mathematics doesn't deal with all of them (not directly, anyway). Here are three ways in which things can be difficult, that maths addresses.

1. Maybe our intuition is not strong enough to work something out.
2. Maybe there's too much ambiguity around making it impossible to work out what's really what.
3. Maybe there are too many problems to sort out and too little time in which to do it.

Mathematics comes to our aid.

1. It helps us to construct and understand arguments that are too difficult for ordinary intuition.
2. It is a way of eliminating ambiguity so that we can know precisely what we're talking about.
3. It cuts corners, answering many questions at the same time by showing that they're all actually the same question.

How does it do it? By abstraction: throwing out the things that cause ambiguity, and ignoring any details that are irrelevant to the question in hand.

You keep doing this throwing-out-and-ignoring, until you get to a point where all you have to do is apply unambiguous logical thought and nothing else.

Bananas and blondes

Ignoring difficult details

Here are some problems that we might try to sort out using our techniques of maths.

1. A banana and a banana and a banana is three bananas, a frog and a frog and a frog is three frogs, and so on. So we think: 'Hmm, something's going on here.' And it becomes $1+1+1=3$.
2. What about if we say: '3 blondes and 2 brunettes is how many people?' We discard the irrelevant notion of hair colour, and the question becomes: '3 people and 2 people is how many people?' And finally this becomes a sum: $3+2=\ ?$
3. My father is twice as old as me but ten years ago he was three times as old as me. How old is he?

   Or: This bag has twice as many apples as that one but if I take ten out of each then this one has three times as many as that one. How many apples are there?
   Both of these become a pair of equations:

$$x = 2y$$
$$x - 10 = 3(y - 10)$$

Now, in this case you might well have been able to do it without explicitly using simultaneous equations, but what about this problem – can you do this one in your head?

> *A rope over the top of a fence has the same length on each side, and weighs one-third of a pound per foot. On one end hangs a monkey holding a banana, and on the other end a weight equal to the weight of the monkey. The banana weighs 2 ounces per inch. The length of the rope in feet is the same as the age of the monkey, and the weight of the monkey in ounces is as much as the age of the monkey's*

mother. The combined ages of the monkey and its mother is 30 years. Half the weight of the monkey plus the weight of the banana is a quarter the sum of the weights of the rope and the weight. The monkey's mother is half as old as the monkey will be when it's three times as old as its mother was when she was half as old as the monkey will be when it's as old as its mother will be when she's four times as old as the monkey was when it was twice as old as its mother was when she was a third as old as the monkey was when it was as old as its mother was when she was three times as old as the monkey was when it was a quarter as old as it is now. How long is the banana?

4. I am very happy. How will I feel if I go bungee-jumping? This has far too much ambiguity. So what does mathematics do with it? It ignores it. (Which makes it much easier.)
5. We want to understand how playing snooker works. So first we imagine that everything is perfectly spherical, perfectly smooth and perfectly rigid. We might think about relevant details like friction, bounciness, spin, and so on later. We can ignore irrelevant details like colour. Except, of course, colour is not irrelevant in practice; but the added pressure of trying to pot the black to win is not a question that mathematics can deal with.

This is the crucial point: we make things easy by ignoring the things that are hard. Mathematics is all the parts we don't have to throw away. The easy bits.

If maths is easy, why is it hard?

You might be wanting to point out a flaw in my argument already: if maths is easy, why does anyone find it hard? There are as many ways to make things difficult as there are to make them easy, and we can be sure that a whole ton of them has been applied to mathematics.

If someone finds maths hard it might also be because nobody told them what it was for – a fork is rather hard to use as a knife. It's also rather hard to use if you're trying to eat a sandwich. Or a bowl of soup. Or a packet of Maltesers.

If someone finds maths hard, it might also be because they have no desire to answer the question that the maths is simplifying. Trigonometry makes triangles really easy. But if you don't care about triangles, you're unlikely to feel that your life has been made easier by trigonometry.

But also, some people just will find things much harder if they're not allowed to use imagination, guesswork or violence. Rationality says that this behaviour is to be deplored as we head towards the ideal of ultimate rationality.

The aim of ultimate rationality

Many people, especially mathematicians, philosophers and scientists, think that we as human beings should aim to become completely rational. That if we discover a way in which we're not rational, we should get rid of it, iron it out, in order to get closer to the goal of ultimate rationality. This has two facets:

1. We should *be* completely rational (that is, behave rationally, think rationally).
2. We should be able to *understand* everything completely rationally.

I want to look at a little logic in order to work out what this might mean.

Background on logic

There's a standard question in logic exams for undergraduates that tries to show, using logic, why democracy doesn't work. This is different from Arrow's theorem that we described earlier, which shows that voting systems can't be fair. This time we're going to show that democracy doesn't work as a policy-making system.

The basic assumption we start with is that everyone in the democracy is *rational*. This is defined in terms of their beliefs: we say that their beliefs should be somehow sensible.

To make it more precise (which is what mathematicians do) we say the beliefs of any individual are 'consistent' and 'deductively closed'. What does this mean?

A set of beliefs is called *consistent* if it doesn't imply a contradiction. For starters this means you don't believe something is both true and false. For example 'I am clever, I am not clever' is clearly inconsistent. But moreoever, you don't believe anything that *leads to* a contradiction. For example, if you believe:

A All mathematicians are clever.

B I am a mathematician.

C I am not clever.

This leads to a contradiction, because A and B together imply that I am clever, which contradicts C.

Your set of beliefs is called closed if anything you can logically deduce from your beliefs is also one of your beliefs. For example, if you believe:

A All mathematicians are clever.

B I am a mathematician.

then you must also believe:

C I am clever.

The exam question then essentially says this: Suppose that there is a vote on all beliefs, and that the government is supposed to act according to what the majority thinks on each belief. Then look at the set of 'things believed by a majority of people' (not necessarily the same majority each time). Is this deductively closed or consistent? The trouble is that it is neither.

Here's how this question looks when written out formally:

> The beliefs of each member i of a finite non-empty set I of individuals are represented by a consistent, deductively closed set S_i of propositional formulae. Show that the set
>
> $$\{t \mid \text{all members of } I \text{ believe } t\}$$
>
> is consistent and deductively closed. Is the set
>
> $$\{t \mid \text{over half members of } I \text{ believe } t\}$$
>
> deductively closed or consistent?

Whether written formally or not, it's all a bit abstract, so let's pick an example. We'll use the following three beliefs:

A University education should be free.

B Everyone should have the chance to go to university.

C We should spend more on universities.

Think for a moment about which of those three statements you agree with. I think you'll agree that if you think university should be free and that everyone should have a chance to go, we'll have to spend more on universities. That is, statements A and B together imply C (unless we allow universities to get a lot worse).

Now suppose we have a grand total of three people in this democracy – we can already produce a problem. Imagine that our three people believe the following things.

* Person 1 believes all three things.
* Person 2 believes university should be free, but we should not spend more on universities. (To make this work, not everyone will be able to go to university.)
* Person 3 believes that everyone should be able to go to university, but we should not spend more on it. (To make this work, university education can't be free any more.)

Now let's see what the majority thinks. In this case, a 'majority' means at least two people.

* Two people believe that university should be free.
* Two people believe that everyone should have a chance at university.
* Two people believe we should not spend more money on education.

Now we try to make policy based on these majority beliefs. We have a problem – we are supposed to make university free and open to everyone, without spending any more money on it. The majority beliefs in this case are neither consistent nor deductively closed. Oh dear.

Life is difficult

Life, frankly, is difficult. And in that context, this idea of a 'completely rational person' is absurd.

The upshot is that rational thinking simply isn't good enough to cope with all that life throws at us. Rationality fails us in life, by being:

* too slow
* too methodical
* too inflexible
* too weak

- too powerful
- and it has no starting point.

And that's why irrationality (or 'arationality') and illogic are not human weaknesses but human *strengths* when used appropriately.

Logic is too slow

In life, we don't always have time to go through logical thought processes to come to a decision. Emergency situations are much more urgent than that, and in that case the important thing is to make a decision that is fast rather than accurate at all costs. There's no point being right if you've already been flattened by the oncoming truck.

How do we know how to throw and catch? How do we sing in tune (if we do sing in tune)? There is maths behind both of those things, but we don't have time to calculate trajectories or vocal cord tensions while catching or singing.

The speed issue is why we have reflex actions. We have built-in reflex actions but we can also train reflex actions, like learning to say 'You're welcome' automatically every time someone says 'Thank you', or learning how to walk all the way to lectures even when you're still pretty much asleep.

Logic is too methodical

Logical thought proceeds calmly step by step through logical inferences. This isn't just slow, it's boring. You don't get into uncharted territory by taking small, safe baby steps. Remember the game of grandmother's footsteps? Someone stands at the front and turns their back. Everyone else stands some distance away and has to try and reach the front first. But the person at the front can turn around at any moment, and if they see you moving, you're sent back to the beginning. My memory of this game is that I never won because I was too cautious; the people who won were the daring ones who took great big steps instead of tiny little ones like me.

The big leaps in life are the flashes of inspiration. These are nothing to do with logic. They happen both in mathematics and in other creative parts of life. The great geniuses of history are often the ones who've made great leaps of inspiration. Now, inspiration in mathematics doesn't mean there's something about mathematics that isn't logical – you still have to use logic to prove what you think is true, but often a flash of inspiration gives you the idea for what you think might be true in the first place.

It's like building bridges: it's hard to build a bridge across a river, but easy to cross the bridge once someone else has built it. And while you're trying to build the bridge, it's helpful to be able to fly.

Logic is too inflexible

Logic is too inflexible in the face of a flexible and often rather random world. Logic is rigid and can't deal with that randomness.

Take our use of language. We assign words to things, essentially producing some random association of sounds with notions. Onomatopoeia aside, there's no logic to it at root. There may be some sense in the etymology of a word, but somewhere back in the history of the word is a random association that started the whole thing off. And we can do that because our brains have the capacity for random association. This is nothing to do with logic.

Logic is too weak

Another situation where logic can't help us is if there isn't enough information. The great thing about logic is that it eliminates the use of imagination and guesswork. But this can be a bad thing too. There are an awful lot of situations in life where we don't have enough information to make a completely logical decision. Perhaps there is an unpredictable element, something random, something we can't detect or things we just don't know, or we haven't got time and resources to find out.

What are we to do, just not make those decisions? Instead, we do various things. We can think about probability. For example, a doctor tells us that 99% of these operations are successful, so we go ahead with it.

We can go instinctive: I don't like the look of this dark alleyway, I'll go a different way. We can guess: like choosing lottery numbers. There's no logic there, but it makes some people exceedingly rich. We can go random ourselves, and let the dice decide.

Decision making is indisputably hard. You try and gather more and more information but at some point your information (or your time) is going to run out, and logic is certainly not going to take you the rest of the way. It's just too weak. Now I'm not saying that you then have to make an irrational decision that actually goes against rationality, but you are going to have to make a non-rational or arational decision. Perhaps if something is pure logic, it doesn't count as a decision at all.

Logic is too powerful

Apart from the fact that logic is too weak, logic is also too powerful. Its unforgivingly brutal power forces us into extreme positions if we take it too seriously.

For example:

It's OK to drink half a pint of beer in an evening.
If it's OK to drink x pints of beer, then it's OK to drink x pints and 1 ml.
In which case, it's OK to drink any number of pints in an evening.

The first two statements seem reasonable by themselves, but the last statement is clearly idiotic. And yet, it follows logically from the first two. It appears that, in order to be rational (closed and consistent), we either have to believe that it's OK to drink any number of pints in an evening (which doesn't sound at all rational) or that it's not OK to drink any beer at all, ever.

The problem here is the subtlety of a fine line, or a sliding scale, or grey area between the black and the white. Somehow, we are able to deal with sliding scales in our heads, in a way that logic can't. The power of logic is in this case its downfall. It brings me to Fuji's paradox.

Fuji's paradox

I've named this paradox after a Japanese bond trader called Fuji who first drew my attention to it. It's a case in point that I don't think he noticed that there was a paradox at all.

It was back in the dark ages, before I realised that mathematics was easy, where bond trading is hard. So there I was, trading futures at Goldman Sachs, when this guy Fuji came along to tell us about the Japanese market. Now, Japanese interest rates were already the lowest in the world, and everyone was wondering whether they'd go any lower, even to zero. Fuji's theory was that they would never actually hit zero because then everyone would know that they couldn't go any lower. Because negative interest rates would be absurd.

The thing is that Japanese interest rates go in increments of quarter percentage points, so the Bank of Japan can only change rates by multiples of that. So, I thought to myself, if Fuji's theory is right then interest rates will not be set at 0.25% either, because then everyone will know it can't go any lower, since it also can't be 0. Oh, but then it can't be 0.5% either. Nor can it be 0.75%, or 1%, . . . , which means that it can't be any percent—which means that Japan can't have interest rates.

This is clearly not true – Japan did have interest rates and still does. So what's gone wrong? (Actually a couple of years later, Japanese interest rates really did go negative, but that's another unbelievable story.)

Unexpected hanging

Fuji's paradox is in fact a manifestation of the 'unexpected hanging' paradox.

The prisoner is told that he will be hanged sometime this week but on a day when he isn't expecting it. So he thinks to himself: Well it can't be on Sunday, because if I hadn't been hanged by Saturday, then I'd know that it had to be Sunday, so I'd be expecting it. So it has to be Saturday at the latest. But then it can't be on Saturday, because if I hadn't been hanged by Friday and Sunday has already been ruled out, then I'd know it was Saturday and I'd be expecting it. So it can't be Saturday and by a similar argument it can't be Friday . . . or Thursday . . . or Wednesday . . . or Tuesday . . . or Monday – which means I won't be hanged!

And then on Monday he is hanged, and he really isn't expecting it.

We can only imagine how miffed he feels, hanging there trying to work out where his logic went wrong.

Logic has no starting point

My last charge against logic is that it has no starting point. If we're not going to take anything on blind faith, we're simply not going to get anywhere. You can't prove something from nothing; you can't deduce anything from nothing; you can't build a Lego construction without any Lego bricks; there's no such thing as a free lunch. We saw Lewis Carroll's paradox, that showed we would at least have to accept the rule of inference *modus ponens* on blind faith, otherwise we would never be able to infer anything from anything else. But even to infer anything from anything else, we have to have something to start with. (Having said that, I've had plenty of arguments with people, mostly mathematicians, who maintain that there's really nothing at all that they believe without justification.)

This seems to me to be an obvious and immediate flaw to the idea of the ultimately rational person. But does that mean we should immediately and completely give up?

The thing is, there's still some scope for greater and lesser rationality. For example:

* A rational person is supposed to believe that the earth is round.
* A rational person is supposed to believe that $1 + 1 = 2$.
* A rational person is not supposed to believe in ghosts.
* A rational person is not supposed to believe in psychic powers.
* Is a rational person supposed to believe in God?

Where do these 'supposed's come from? They come from society. It hasn't always been the norm to believe that the earth is round. And in some societies it is the rational norm to believe in God and in others it is not. So, in fact, rationality is a *sociological* notion. Apparently you can still count as rational as long as all your basic beliefs come from the big bank of basic beliefs accepted by society as 'rational things to believe'. If your basic beliefs are 'the moon is made of soft green cheese' and 'sleeping upside down is good for the elbows' or 'I must kill as many people as possible', then someone will soon come and take you away.

But still, I've had arguments with people (mainly philosophers) who get very upset if something I'm defending comes down to something I believe, and I declare that I believe it 'because I do'. Rational people aren't supposed to do that are they?

Well I believe it's a good thing to be aware of what you're assuming. I repeat: I *believe* it's a good thing to be aware of what you're assuming. Whether it's a whole lot of things at the root of your belief tree or, say, God.

Being aware of your assumptions is definitely part of the discipline of mathematics, and also part of what makes maths easy – everyone has to state very clearly what their basic assumptions are. I don't think there's anything wrong with believing some things without justification – they are your axioms, from which all else grows. For example, I believe in love, but I have no justification for that. The crucial part is to be aware that it's one of your axioms, and not to pretend you arrived at it by logic.

Mathematics is not life

So: maths is easy, life is hard, therefore maths isn't life.

This doesn't mean that we shouldn't try to extend the scope of mathematics for it to include as much as possible, just as we should try to become 'more and more rational' by continuing to work out what the inital premises of our beliefs are. The pursuit of mathematics is 'the process of working out exactly what is easy, and the process of making as many things easy as possible'.

But we should not feel affronted by the existence of things that can't be subsumed by mathematics, the irrational or arational, the illogic. Without that, there would be no language, no communication, no poetry, no art, no fun.

part two

CATEGORY THEORY

9 WHAT IS CATEGORY THEORY?

Not much mathematics was needed before people started doing trade. Numbers themselves weren't even necessary, let alone the more complicated things you can do with them. Negative numbers don't make much sense if you haven't thought about the possibility of going into debt.

Children don't really need numbers in the early part of their lives. If we deliberately teach them numbers then they have the capacity to pick them up when they're one or two, but if we don't actively teach them the concept, I'm not sure when they'd pick it up. Plenty of children arrive at primary school at the age of five, being able to recite their 'number poem' without being able to use it to count anything. In everyday adult life it's hard to avoid numbers even if it's nothing other than prices at the supermarket, but small children get by just fine without numbers.

Likewise, mathematics got by just fine without category theory for thousands of years, but now, in everyday mathematical life it's hard to avoid it – at least in *pure* maths.

The distinction between 'pure maths' and 'applied maths' is a bit spurious, or at least the grey area where they meet is pretty grey and quite large. But broadly speaking applied maths is a bit closer to normal life. Applied maths is more likely to be modelling real things in life like the sun, water flowing through pipes, traffic flow. It could be thought of as the theory behind things in real life.

Pure maths is one step more abstract: it is the theory behind applied maths. This is a simplification, but it will do for now.

Lego, yet again

The difference between pure and applied Lego

Are you more interested in using basic Lego bricks to build fantastic big sculptures? Or are you be interested in buying all the complicatedly designed little pieces to build machines, or working robots, or train sets, or spaceships? Even if you don't do Lego yourself, what do you find more fascinating – a Lego version of the Eiffel Tower built only from basic 2-by-4 pieces, or a fantastic articulated robot built from complex high-tech pieces? Using special pieces will be quicker, and you'll get a more realistic model. For example, you can have real wheels with tyres, instead of sort of bumpy angular ones. But there's something immensely satisfying and impressive about having whole buildings and towns built from basic pieces. The creativity and ingenuity required to do it are fascinating.

Pure maths is like using only the basic Lego bricks and building everything from scratch. Applied maths is like using special pieces. Applied maths more closely models real life, but pure maths is at the heart of it, just as you can't get away from the 'pure' Lego-building techniques just because you've acquired wheel pieces.

Topology is a part of pure mathematics that studies the shapes of things like surfaces. We've talked about how topology studies which shapes can be morphed into other shapes without breaking them or sticking them together, but it also studies what happens when you *do* cut them up and stick them together, and how you can build more complicated shapes from simpler ones. It is in fact quite a lot like Lego.

Topology gets used in quantum mechanics, to build models of subatomic particle behaviour. This is called 'topological quantum field theory' and is probably somewhere in the grey area between applied mathematics and theoretical physics. A more large-scale part of applied topology is in cosmology, where the shape of space-time is studied.

Even further along the applied scale is where topology is applied to the study of knots in DNA and configurations that robotic arms can get into. This takes us into biology and engineering.

Another example along the scale of pureness comes from calculus. At root, calculus is the study of infinitesimally small things, or things that are changing continuously rather than in jumps. This is an important area of pure mathematics. As a field of pure study, it is concerned with things like whether a quantity is changing smoothly, and what its rate of change is.

It leads to the question of solving equations involving quantities *and* their rate of change at the same time. For example, if something is moving, we might know about the force being applied to it and the speed it is going, as well as the position it is in. This sort of equation is called a *differential equation* and this takes us further towards applied mathematics than pure. It relates to things like gravitational pull, radioactive decay and fluid flow.

When these things get applied to specific real-world situations, we have gone out of the realm of applied mathematics and right into engineering or medicine or even finance. Differential equations are one of the most widely applied pieces of mathematics all over the place, as almost all measurements of things in real life are somehow fluctuating at some rate or other.

Lego Lego

When it is possible to build things out of themselves

Have you ever tried making a Lego brick . . . out of Lego? It would be a sort of meta-Lego brick. Instead of a Lego train or a Lego car or a Lego house, you'd have built 'Lego Lego'. I have seen pictures of cakes made out of Lego bricks – a Lego cake. And I've seen Lego bricks made out of cake: cake Lego.

And, inevitably, there are cakes made out of Lego bricks that are themselves made out of cake: cake Lego cake.

Category theory is the mathematics of mathematics, a sort of 'meta-mathematics', like Lego Lego. Whatever mathematics does for the world, category theory does it for mathematics. This means that category theory is closely related to logic. Logic is the study of the reasoning that holds mathematics together. Category theory is the study of the structures that hold mathematics up.

At the end of the last chapter I suggested that mathematics is 'the process of working out exactly what is easy, and the process of making as many things easy as possible'. Category theory, then, is:

> *The process of working out exactly which parts of maths are easy, and the process of making as many parts of maths easy as possible.*

In order to understand this we need to know what 'easy' means inside the context of mathematics. That's really at the heart of the matter, and is what we'll be investigating in this second part of the book. In the first part we saw that mathematics works by abstraction, that it seeks to study the principles and processes behind things, and that it seeks to axiomatise and generalise those things.

We will now see that category theory does the same thing, but entirely inside the mathematical world. It works by abstraction of *mathematical* things, it seeks to study the principles and processes behind *mathematics*, and it seeks to axiomatise and generalise those things.

Mathematics is, if you like, an organising principle. Category theory is also an organising principle, just one that operates *inside* the world of mathematics. It serves to organise mathematics. Just as you don't need a classification system for your books until you have quite a vast collection, mathematics didn't need this kind of organising until the middle of the twentieth century, which is when category theory grew up. Systematising things can be time-consuming and complicated, but the idea is that in the end it's supposed to help you think more clearly.

Category theory is the study of the mathematical notion of 'categories'. Although this is a word taken from normal life, it has a different and carefully formulated meaning in mathematics. These mathematical things called categories were first introduced by Samuel Eilenberg and Saunders Mac Lane in the 1940s. They were studying algebraic topology, which turns shapes and surfaces into pieces of algebra in order to study them more rigorously. Originally this involved relating all those shapes to *groups*, the notion that we introduced and axiomatised in the previous part of the book. They realised that in order to keep a clear head while doing this, a more powerful and expressive type of algebra was needed, a bit like groups but with some further subtleties. Mathematics had become vast enough to need its own system of organisation. Mathematics needed to think more clearly. And so category theory was born.

Then something wonderful happened. Just as mathematics began as the study of numbers, but then people realised the same techniques could be used to study all sorts of other things, category theory began as a study of topology, but then mathematicians rapidly realised that the same techniques could be used across huge swathes of mathematics. Category theory grew up to have greater influence than its 'parents' ever imagined.

10 CONTEXT

Lasagne

Ingredients

Bolognese sauce

Fresh lasagne sheets

Béchamel sauce

Grated parmesan

Method

1. Spread a layer of bolognese sauce over the bottom of a baking dish. Cover with a layer of lasagne sheets, then a layer of béchamel sauce.
2. Repeat twice more, finishing with a layer of béchamel sauce.
3. Sprinkle parmesan over the top and bake at 180°C for 45 minutes or until it looks delicious.

When you see this recipe, you might think 'Lasagne – that's easy.' Or you might think 'Béchamel sauce? How do I make that?' This recipe is very simple, but only because it assumes that you already know how to make bolognese sauce, béchamel sauce and pasta. If this were a recipe with instructions from scratch, it wouldn't be simple at all – it would have a long list of ingredients and many steps.

Recipes look very different depending on what sort of chef they're aimed at. Is it for an experienced professional? A serious amateur? A novice still learning basic skills? Category theory emphasises the *context* in which we're thinking about things, rather than just the things themselves. This includes what sorts of details we're interested in right now, what features do and don't matter in this situation, what counts as a basic assumption, and what needs to be broken down further. Just like in the lasagne recipe, where béchamel sauce counted as 'basic', there are some situations where the number 5 counts as basic, and others where it doesn't. In the context of just the natural numbers (1, 2, 3, 4, 5, 6, and so on), the number 5 some very particular characteristics: it can only be divided by 1 and 5, which is to say it is a prime number. However, in the context of rational numbers (fractions) it can be divided by all sorts of things. 5 divided by 10 is $\frac{1}{2}$, for example. 5 divided by 2 is $2\frac{1}{2}$. The character of the number depends on the context in which we've placed it.

Brothers

Putting people in context by finding out about their family

I met a guy at a party recently who, after a little bit of conversation, said to me 'Do you have brothers? I bet you have brothers.' I said no, and asked him why he thought I must have brothers. He replied, 'Because you're not afraid of talking to tall, handsome men.'

Another guy at another party told me 'You're so self-sufficient, I bet you're an only child.' Also wrong, but it brought to mind one of my favourite scenes in Casino Royale, where James Bond and Vesper Lynd meet for the first time and verbally spar with one another on the train. Bond coolly declares she must be an orphan, and she, equally coolly, surmises that 'Since your first thought ran to orphan, that's what I'd say you are.'

Indeed, I suspected the guy who thought I was an only child was one himself and, fancying myself as Vesper Lynd, that's what I asked him. It was true.

It's natural when getting to know someone to be interested

in their family, their childhood, where they are from. Some people think these questions are boring and pointless, or perhaps they are miffed by the questions because they feel that these basic facts about themselves do not give a very accurate impression of what sort of person they now are.

However, it is all part of the process of understanding a person in some sort of *context*, rather than in isolation. One of the things that makes us human is the way in which we interact with other humans. An autobiography of a famous person would not be very interesting if it did not contain any description of the person's family, friends and relationships. An absolute character study, out of context of other humans, would be almost impossible to achieve.

In the same way, category theory seeks to emphasise the context in which things are studied rather than the the absolute characteristics of the things themselves.

This is just like we did with the 'lattice' of factors of 30, where just writing the factors in a list is not nearly as interesting as drawing a picture showing how they are related to one another:

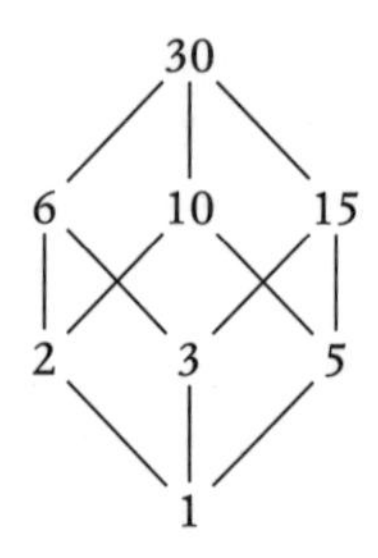

This is a way of putting the factors into *context* and in the next chapter we'll see how exploring the relationships between things is a good way of doing that.

If you remember what highest common factors and lowest common multiples are, you might notice some patterns in that picture relating numbers in one row and the numbers connected to it in the row above and below.

Mathematicians

Putting people into context by finding out what they do

I once went to a party and decided to try an experiment: I refused to tell anyone what my job was. Telling people you're a mathematician produces all sorts of odd responses. Some people become afraid, and extract themselves very quickly, but others immediately start trying to demonstrate how 'intelligent' they are. Yet others immediately start trying to belittle me. One guy responded, 'But what are you going to do after that?', to which I replied, of course, that I wanted to be a mathematician for life.

The absurd conversation proceeded like this:

Him: Well, you'll never get a job.
Me: Actually I've already got a job.
Him: Well, you'll never get a permanent job.
Me: Actually I've already got a permanent job.
Him: What, in a school or something?
Me: As a university lecturer.

Another guy discovered I was a mathematician and started grilling me on my credibility.

Him: You mean, you work for a bank?
Me: No, I work for a university.
Him: Just teaching?
Me: I do teaching and research.
Him: Do you have a PhD?
Me: Yes.
Him: Where did you get it?
Me: Cambridge.
Him: Oh, PhDs in England are so easy to get, they don't really count.

I channelled Vesper Lynd and surmised that this guy must be a failed mathematician: it turned out, he had failed to get a PhD place in France and had gone to work for a bank instead. The first guy turned out to be a maths teacher. In a school.

On another occasion, someone blurted out 'You mean, like the book *Categories for the Working Mathematician*?' As it happens, I was just starting to study category theory at the time, and desperately trying to get hold of a copy of this crucial book, written by one of the founders of category theory, Saunders Mac Lane. But it was out of print and I couldn't find one anywhere. This guy happened to own a copy, which he had used some years earlier when he had been a student, but he was no longer in mathematics and promised to send me his copy of the book.

So I am happy to report that sometimes there are advantages to putting myself in context.

Sometimes a mathematical object has several jobs, and one of them will give us a more illuminating context than others. This is just like when a person has two jobs, one of which tells us more about their personality than the other. Perhaps they're an office manager and also a salsa teacher.

Here's a mathematical example. The number 1 can be characterised by its 'job' as a multiplicative identity. This means that whenever you multiply another number by 1, nothing happens. However, this doesn't tell us much what context we're thinking about, because it's true no matter what sort of numbers we're dealing with.

The number 1 has another 'job', which is that if you keep adding it to itself, you get *all* the natural numbers 1, 2, 3, 4, 5, . . . In mathematical language we say 1 *generates* the natural numbers. This job is very much tied to the context of the natural numbers.

Online dating

Understanding people by seeing them in different contexts

When you have a new partner, it's always a big moment when you first meet their friends – unless you already knew their friends. With the proliferation of online dating, this is becoming a much bigger issue. Meeting online is like meeting completely out of context. It's not like meeting through mutual friends, or shared interests, or shared experiences. This can also be true if you meet someone at work, and there's a certain moment when you first see them in the context of their non-work friends.

People can turn out to be very different in different contexts. It's normal for people to be different at work and outside of work, even if it's just that they're more reserved at work and let their hair down more outside work. For most of my career I've been much less myself at work, for fear of drawing too much attention to the fact that I'm female in an extremely male-dominated environment. I tried to be as unfemale as possible, to avoid the accusation that my being female was making me a worse mathematician.

But also people can be quite different among different groups of friends. Some people you're friends with out of *longevity* – you grew up with them, and that long shared experience will keep you together even if on the face of it you no longer have that much in common. Lives and people diverge after all.

There are people you're friends with out of *proximity* – they happen to be around in your normal life. Perhaps you see them every day at work, perhaps they're your neighbours, perhaps you see them at the gym or in your salsa classes or your book club, or they have children who are friends with your children, or you take the same bus to work with them every day. I've made several friends on trains.

But then there are the people you're friends with out of *affinity*. You have something in common with them that isn't circumstantial, but is something deep in your character. I have deep friendships with many category theorists around the world despite the fact that we've never lived in the same city, country or in some cases hemisphere.

Anyway the point I'm getting to is that you might well behave differently among these different types of friends. You might talk about different things, discuss them in different ways, meet them in different types of places. So who is the 'real' you? Is it who you are among your family? And yet, many of us revert to being like small children with our family, uncovering old frustrations and perhaps slipping back into the roles we had when we were growing up. It is hard to break out of those roles.

Or is it who you are among your 'affinity' friends? This is a bit like the question of whether you're more you or less you when you're drunk and saying things you perhaps wouldn't say if you were sober. Are those things more honest or simply more extreme?

Category theory does not try to answer the question of which is 'more real'. We study the number 5 in the context of whole numbers, and in the context of fractions, but we do not pass judgment on which *really* is the number 5.

* In the context of natural numbers $(1, 2, 3, 4, \ldots)$, 5 is a prime number, that is, it is only divisible by 1 and itself (and is not equal to 1). It does not have an additive inverse or a multiplicative inverse.
* In the context of integers $(\ldots, -3, -2, -1, 0, 1, 2, 3, \ldots)$, 5 now has an additive inverse, which is -5. That is, if you add 5 and -5 you get the additive identity 0. But 5 does not have a multiplicative inverse.
* In the context of rational numbers (fractions), 5 has a multiplicative inverse, which is $\frac{1}{5}$. That is, if you multiply 5 and $\frac{1}{5}$ you get the mutliplicative identity 1. And 5 isn't prime any more because it is divisible by all sorts of things. For example 5 is divisible by $\frac{1}{2}$.
* In the context of arithmetic on a six-hour clock ('modulo

6') 5 is actually a *generator* for the number system. This means if you add 5 to itself repeatedly you will eventually get every number in the system. You can try it, remembering that the only numbers in this system are $0, 1, 2, 3, 4, 5$, and every time you get to 6 you count it as 0 again. So $5 + 5 = 10$, which is the same as 4. $4 + 5 = 9$, which is the same as 3. If you keep going, adding 5 repeatedly, you'll get 2 next, then 1, then 0, showing that 5 really does generate all the numbers. By contrast, 5 definitely doesn't generate all the natural numbers, because if you keep adding 5 to itself you'll get $5, 10, 15, \ldots$ and only ever achieve the multiples of 5.

So we see that the number 5 takes on different characteristics depending on what context we're in. Category theory seeks to highlight the context you're thinking about at that moment, to emphasise its importance and raise our awareness of it. The way it does it, as we'll see in the next chapter, is by emphasising relationships between things rather than just their intrinsic characteristics. Because as we've seen, even for something as simple as the number 5, the 'intrinsic characteristics' aren't so intrinsic after all.

You might wonder which other numbers are generators for the six-hour clock. Of course 1 will work, but if we try adding 2 to itself repeatedly, we'll get 2, 4, 0, 2, 4, 0, . . ., so we'll never hit any of the odd numbers.

For 3, we'll get

$$3, 0, 3, 0, 3, 0, \ldots$$

and for 4, we'll get

$$4, 2, 0, 4, 2, 0, \ldots$$

so 3 and 4 aren't generators either. So being a generator is a fairly special characteristic.

Arsenal

When things are more exciting in one context than in another

People often look extremely different out of context. Conductors, for example, often turn out to be much shorter than I imagine, because you only ever see them standing up, and on a raised podium, and in a position of massive authority. Students often turn out to be much taller than I imagine, because most of the time I see them they're sitting down and I'm standing up, and I'm the one in the position of authority.

I was once in a bar in London when Arsenal walked in – the whole team and the entire entourage. I was sitting there doing some maths, as I sometimes do in bars, because I like being surrounded by people and I like the feeling that I'm surrounded by people having fun.

Anyway I was sitting at this bar with my pen and my black notebook in which I write every single thought, and this huge crowd of people wearing football shirts came in. Being rather football-ignorant myself, I didn't recognise the shirt, but watched these young, lanky, slightly awkward-looking, mostly Mediterranean youngsters file in with some older guys who were clearly their minders. The young ones went straight up to their rooms, and the older guys came into the bar, and I thought, 'Oh, it must be some sort of visiting youth football team from Europe. Lucky them that they get to stay in such a swanky hotel!'

I didn't think much of it, and carried on doing my maths, until one of them came to the bar and struck up a conversation with me.

'Is that chemistry you're doing?' he asked, peering at my notebook. I explained that I was a mathematician, and realised that he was now close enough for me to read what his shirt said: Arsenal.

Now you might think I'm a bit thick at this point not to realise it was Arsenal in front of my nose, but after all, people go round wearing David Beckham shirts and it doesn't mean they're David Beckham. So then I uttered the immortal line, 'Um, are you part of some sort of... team?'

'Yes, it's called Arsenal.' the guy replied, kindly. And then he added 'It's a Premier League football team.'

I suddenly thought back to those lanky youths sloping obediently up the stairs to their rooms. They're all millionaires, and famous! They were very out of context.

In maths there are also concepts that are pretty unexciting in one context and extremely exciting in another. A good example is the Möbius strip, which is made from a strip of paper by sticking the ends together, but instead of making a normal cylinder:

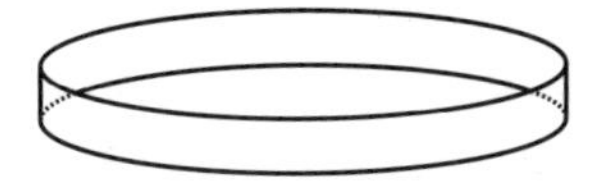

you twist the paper before sticking the ends together:

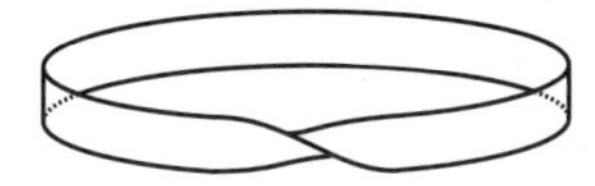

This is a very exciting surface, because it *only has one side*. You can try making one of these, and try colouring in one side. You will discover that you get all the way round and keep going, and you will go all the way round again and get back to where you started, having coloured in what looked like both 'sides', but without ever taking your pen off the paper. This is quite exciting. Better still, you can try cutting one out of a bagel, and then spreading cream cheese on it. You will discover that if you try and spread cream cheese on just one side, you will actually end up covering both 'sides' because in fact there is only one side.

However, from the topological (playdough) point of view, a Möbius strip is not that interesting because it's 'the same' as a circle: if you started with an ordinary circle (ring shape) of plasticine or playdough, you could make a Möbius strip just by squashing it around, without making new holes or sticking things together. You would have to flatten the playdough out working your way around the circle bit by bit, twisting your flattening action as you go. (This is a bit hard to imagine, so you might want to try it; if you don't have any playdough to hand, you could make some basic dough out of roughly equal quantities of flour and water.) It turns out that the Möbius strip is an interesting *tool* in topology, but is not interesting by itself.

The way this is technically stated is by appealing to the different notions of sameness that go best with different contexts. The notion of sameness that we've introduced for topology is the playdough type, which is called *homotopy equivalence*. So technically we say that the Möbius strip is homotopy equivalent to a circle. This is useful, but unsatisfactory, because a Möbius strip is much more exciting than a mere circle.

One way this can be expressed is by a more sophisticated mathematical structure called a 'vector bundle'. Remember earlier on when we were imagining a magic pen that could draw in mid-air? Imagine if they then invented a thick pen as well – like the kind you can use on paper to draw a line with serious width, because the nib itself is in the shape of a straight line. Imagine one of those that could draw in mid-air – you'd be able to draw whole surfaces in mid-air. That would be amazing. It would be like waving a light sabre around, except leaving a track.

Now, if you imagine drawing a circle in the air with a light sabre, the surface you make is a 'vector bundle over a circle'. The idea is that for each point in the circle, you now have an entire vector, that is, a line given by the light sabre at that instant.

The thing is that while you're drawing the circle in the air, you can twist the light sabre around if you want. So perhaps you're drawing a circle by running round in a circle. At the beginning and end you have the light sabre vertical, so that your thickened circle does meet up. If you just hold the light sabre pointing straight up the entire time, you'll draw a cyclinder in the air. But what if you start with the light sabre pointing to the sky, and as you run round, you gradually bring it down until at the end it's pointing to the floor.[1] In that case you'll have drawn a Möbius strip, although you yourself still only ran around in a circle to draw it. The *topology* of the situation only notices that you ran round in a circle both times, so it can't tell the difference. But the *vector bundle* structure notices what twisting you were doing with your arm at the same time, so it does see the difference.

Think of a number

Here is a basic example of how you can find out all about something just by looking at its relationships with other things.

> *I am thinking of a number. If I add 2 to my number, I get 8. What is my number?*

Well, it's not very hard to work out that the answer here is 6 (my favourite number). Let's try this one.

> *I am thinking of a number.*

1. *It is a positive number.*
2. *If I subtract 8, the answer is negative.*
3. *If I divide by 3, the answer is a whole number.*
4. *If I add it to itself, the answer has two digits.*

> *What is my number?*

[1] The key here is that your hand doesn't stay at the same height as you run round – the centre of the light sabre has to stay at the same height. Perhaps we should use Darth Maul's double-ended light sabre so we can hold it in the middle.

Yes, the answer is still 6. Not very original, am I? However, the point wasn't originality – the point, as usual, was to make a point: you can understand something via its relationships with other things. The examples involving my favourite number were silly examples designed for nothing other than to make a point. (This is the kind of example that can make people think maths is useless. But some examples aren't there to be 'useful' as much as 'illustrative'.)

The point of that example was that category theory elevates the importance of relationships, so that it becomes perhaps even more important than studying intrinsic properties of things.

One basic example is the idea of a number line. The important thing about the numbers 1, 2, 3, . . . is not what they're called, but what *order* they go in. It doesn't really matter what they're called, as long as the words (or symbols) always go in the same order. So it's sensible to draw them in a line:

$$\ldots -4 \quad -3 \quad -2 \quad -1 \quad 0 \quad 1 \quad 2 \quad 3 \quad 4 \ldots$$

What this really does is emphasise the relationship between them, and keep them fixed in their positions. There are various ways to generalise this. If we allow all real numbers (rational and irrational) we get to fill in all the spaces between 1, 2, 3, . . . and the line goes on 'forever' in both directions. We can't physically draw that, but we can imagine it:

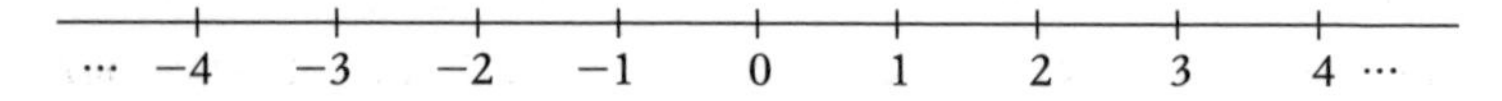

Now let's think about the imaginary numbers we introduced in Chapter 6, with i being $\sqrt{-1}$, and then the multiples $2i$, $3i$, $4i$, and so on. These will also be in a line, and in fact we can imagine ai where a is any real number, so that we can fill in the gaps in this line. However, this line should not be confused with the line for real numbers, as it's completely different. So we often draw it vertically instead of horizontally:

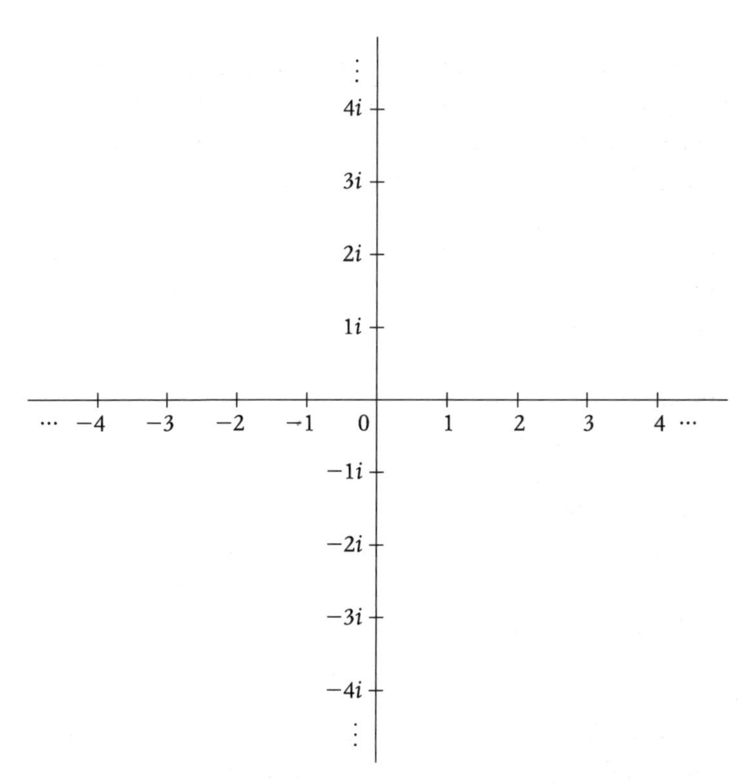

You might naturally now start wondering what happens in the space surrounding those two lines. And we arrive at the same answer as if we ask the following question: can we add and multiply imaginary numbers according to the axioms for a *group*? Adding them is fine, because we get things like $2i + 3i = 5i$. Because i should behave just like apples, monkeys, or anything else: 2 of them added to 3 of them gives 5 of them.

But what if we try multiplying them? We already know that $i \times i = -1$, which is *not* an imaginary number. So we have a problem. What about something like $2i \times 2i$? If we assume the usual laws of multiplication, we should be able to say

$$\begin{aligned} 2i \times 2i &= 2 \times i \times 2 \times i \\ &= 2 \times 2 \times i \times i \\ &= 4 \times (-1) \\ &= -4 \end{aligned}$$

and similar things. We could write this abstractly and say that if a and b are any real numbers, then

$$ai \times bi = -ab.$$

In any case, an imaginary number times an imaginary number will always be a real number. This is a bit like the rule saying a negative number times a negative number is positive, whereas a negative times a positive is still negative. Similarly an imaginary number times a real number is still imaginary. We can sum this up in some tables:

×	positive	negative
positive	positive	negative
negative	negative	positive

×	real	imaginary
real	real	imaginary
imaginary	imaginary	real

Now we have a problem – or just an interesting issue. Because if we want to be able to add *and* multiply, we are going to have to mix up the real and imaginary numbers. For example, what if we want to do

$$2i \times 2i + 2i?$$

We know that $2i \times 2i = -4$, so $2i \times 2i + 2i$ should really be $-4 + 2i$. What is that? We have invented the *complex numbers*. These are what you get when you allow yourself to add up real numbers together with imaginary numbers. And this is what fills in the 'space' around our real number line and our imaginary number line:

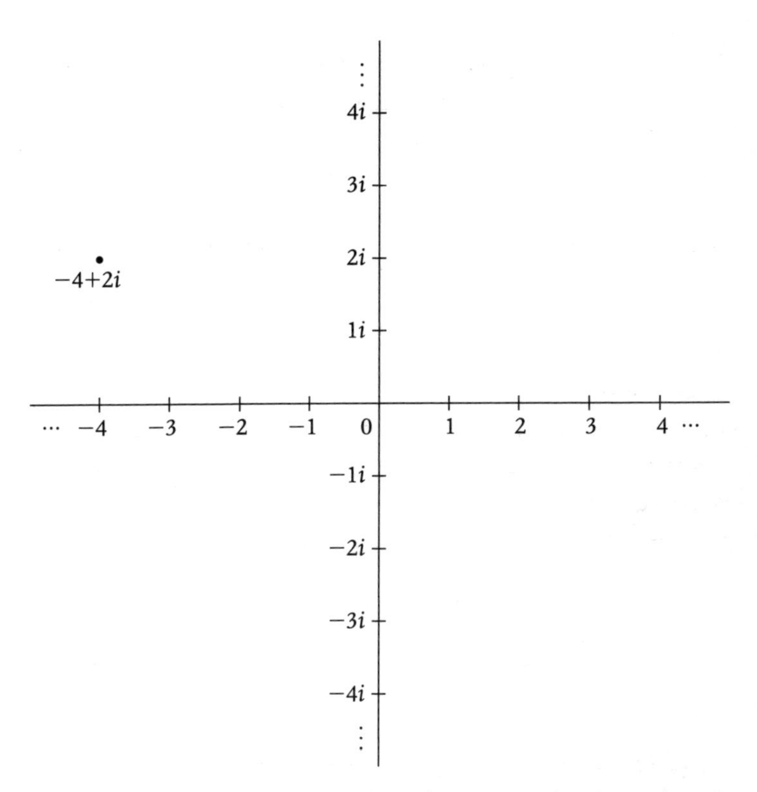

This is like a map where everything has an x and a y coordinate, except now everything has a 'real' coordinate and an 'imaginary' coordinate. So the point with coordinates (x, y) is the complex number $x + yi$. This might sound a bit abstract – because what *are* these things? Whatever they 'are', we can add and multiply them just like real numbers; moreover, we now have solutions to *all* quadratic equations, even though the equations themselves only have real numbers in them. We've already seen that the equation

$$x^2 + 1 = 0$$

now has a solution. In fact is has two solutions: i and $-i$, because by the usual rules of multiply negatives, $-i \times -i = i \times i = -1$. So just like all other numbers (apart from 0), -1 also has two square roots: i and $-i$.

Now *every* quadratic equation has a solution. For example, the innocuous looking equation

$$x^2 - 2x + 2$$

couldn't be solved just using real numbers, but using *complex* numbers we get two solutions $1 + i$ and $1 - i$.

You can just substitute the numbers in and try it, as long as you keep a clear head about how to multiply complex numbers. You just multiply out the brackets slowly. We can try it with $x = 1 + i$:

$$\begin{aligned}(1+i)^2 - 2(1+i) + 2 &= 1(1+i) + i(1+i) - 2 - 2i + 2 \\ &= 1 + i + i + (-1) - 2 - 2i + 2 \\ &= 0\end{aligned}$$

because all the i's cancel out, and all the real numbers cancel out. You can try it for $x = 1 - i$ as well.

Complex numbers are such abstract things that it can be very hard to get your head around them at all. They really only exist because we imagined them. But in a way this is no different from a perfect circle or a straight line – these things are all in our heads only, and don't exactly exist in 'real' life. Remember, in maths anything exists if you can imagine it, and it doesn't cause a contradiction. Representing complex numbers in this grid with real numbers puts them in a useful context. It gives us a way of thinking about them that relates them to each other, and relates them to things that *do* exist in real life – two-dimensional patterns – so it helps us give these abstract things their meaning. Category theory also turns the relationships between things into patterns that we can draw on a page, as we'll see later.

We're going to see that category theory works by picking what relationships between things we are interested in, and emphasising those. We'll even *generalise* the notion of relationship to encompass things that at first sight didn't look very much like relationships, so that we can study more and more situations using the same way of thinking. This is the subject of the next chapter.

11 RELATIONSHIPS

Porridge

Ingredients

1 cup of oats

2 cups of water

Salt to taste

Method

1. Put all the ingredients in a pan and bring to the boil.
2. Reduce the heat and stir until done to taste.

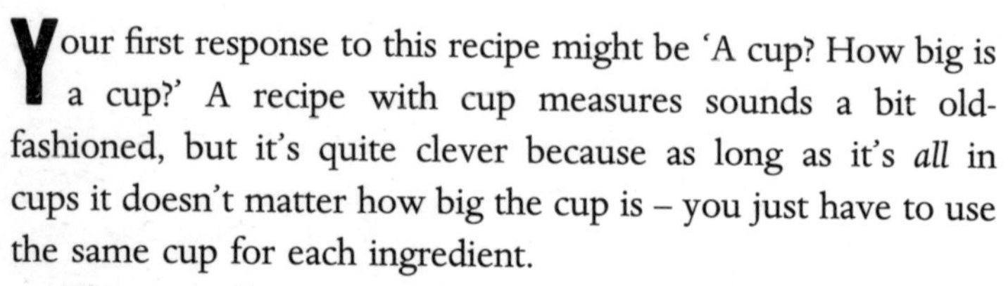

Your first response to this recipe might be 'A cup? How big is a cup?' A recipe with cup measures sounds a bit old-fashioned, but it's quite clever because as long as it's *all* in cups it doesn't matter how big the cup is – you just have to use the same cup for each ingredient.

This sort of recipe emphasises the *relationship* between the things in the recipe, rather than their absolute quantities. This is what category theory does as well. Instead of just studying objects and their characteristics by themselves, it emphasises their relationships with other objects, as the main way of placing them in context.

Feminism

When equality isn't equality

You might typically think of maths in terms of numbers and equations. So far I have described various mathematical objects that aren't numbers, and now it's time to think about things that aren't equations either. What on earth would an equation involving circles mean? Or an equation involving surfaces or spheres?

The most straightforward relationship between things is equality. But equality in mathematics is a more stringent notion than 'equality' in normal language. When we talk about 'equality' in normal life, we usually mean equality just from some point of view. If you think men and women are equal, I doubt you think they're actually exactly the same. You probably mean they contribute just as much to society as each other, and deserve to be treated just as well as each other by society. We can handle this sort of interpretation in normal language – just about. After all, there are still plenty of arguments about exactly what 'equality' means, socially. However, in maths we certainly can't handle this sort of haziness. We are only supposed to reason using hard logic, not subjective interpretations of things. According to hard logic, two things are only equal if they are exactly, precisely the same in every way. In maths nothing is equal to me except me.

You might think that this is an annoying piece of pedantry, and perhaps it is. Sometimes the quest to rule out ambiguity can lead to this sort of annoyance, where something that used to have meaning becomes so unambiguous as to lose almost all its meaning. You might be tempted to throw up your arms in frustration and give up at this point. In fact, maybe you *did* throw up your arms in frustration at exactly this sort of thing, and that's why you're not a mathematician (if you're not). But mathematics doesn't give up at this point. Mathematics says: fine, that was just the first step. We proceed in baby steps. With each step we get a bit closer to what you *really* meant with some other notion that can also be made unambiguous.

In category theory this means thinking about some broader types of structure of which equality is just one example. This would allow some other types of relationships to exist, other than this excessively restrictive notion of 'equality'. We have already seen some examples of things that are more or less 'the same' in some contexts. For example, similar triangles are not precisely the same as each other, but close. Then there's the idea of 'the same' we thought about for doughnuts and coffee cups. And what about the relationship between the symmetries of an equilateral triangle and the different ways of ordering the numbers 1, 2, 3? We will look more specifically at different notions of 'sameness' in a later chapter, but for now we'll look at relationships in general, whether they're sameness or not. Category theory has an eventual aim of expressing interesting notions of sameness, but it starts by focusing on relationships in general.

The relationships are actually called 'morphisms' to allow for the fact that they might not be quite like relationships. For example:

- A matrix with two rows and three columns can be very usefully thought of as a morphism from 2 to 3, but it's a bit tenuous to think of it as a relationship between the numbers 2 and 3.
- We will see that object can have many different morphisms to itself, but it's a bit harder to think of an object as having many different relationships with itself.

Sometimes we name mathematical concepts using words from everyday life to appeal to our intuition, but sometimes we invent words in order to try *not* to be biased or limited by our intuition.

Here are some examples of words from everyday life that have been appropriated by mathematicians to mean something technical: root, prime, rational, real, imaginary, complex, biased, natural, weighted, filtered, category, ring, group, field.

Here are some examples of mathematical words that are not really from everyday life, or have simply been invented: logarithm, surd, morphism, functor, monoid, tensor, torsor, operad.

Here are some relationships that category theory looks at:

* Whether numbers are greater than one another.
* Whether numbers divide one another.
* Whether spaces can be deformed into one another in the manner of playdough.
* Functions from one set to another. A function is a process that takes things in one set as an input and produces something in the other set as the output. Note that we can have many different functions between the same two sets, producing different outputs. This is why in the end we need to think about not just *whether* things are related, but also *how* they're related.
* A good notion of relationship between groups is a *function* that also interacts sensibly with the way of combining objects in the group. We'll come back to this later.

Erdős number

Measuring all relationships relative to one very special person

The theory of 'degrees of separation' of human beings is about how long a chain of acquaintances you have to go along to get from any human being to any other. For example, everyone I know is separated from me by only one step, but their acquaintances are separated from me by two steps (unless I know them already). The theory is that it only takes six degrees of separation to link any two people in the world.

One interesting thing to do is to replace 'acquaintance' with 'co-author'. So if I have published a paper written jointly by me and someone else, we are one mathematical step apart. You can then draw diagrams of these relationships and wonder how many degrees of separation it takes to reach all mathematicians in the world.

Paul Erdős was an eccentric Hungarian mathematician of

the twentieth century. He was eccentric even in the context of mathematicians – he had few possessions, lived a nomadic life travelling from place to place with just the suitcase containing his possessions, and fuelled his mathematics with coffee and amphetamines.

He was also a prolific collaborator, in fact, possibly the most prolific of all time: he published papers with 511 different co-authors in his lifetime. (By contrast, I have six so far.)

His friends came up with the idea of linking all mathematicians to Erdős by degrees of separation. So all his collaborators are one step from him, their collaborators are two steps from him (unless they've also actually collaborated with him), and so on. The degree of separation is light-heartedly called the Erdős number. So his 511 co-authors have Erdős number 1, and there are about 7000 people with Erdős number 2 – including me. By the time you get to six degrees of separation, this encompasses 250 000 people. They are not all mathematicians – it branches out into statistics, astronomy and genetics as well, among other things.

This relates to an important concept in category theory. Once you've decided what kind of relationships you're going to focus on, you can wonder whether there's one 'special object' in your world that somehow encapsulates tons of important information all by itself. That is, a sort of barometer object, a litmus test object, a benchmark object, an Erdős-like figure. Mathematicians call this a *universal property*.

Defining important things by their relationship with other things is something that we have already discussed.

* There is the number 0, which is the only number with the property that when you add it to other numbers, nothing happens.
* There is the number 1, which is the only number with the property that when you multiply other numbers by it, nothing happens.

- There is the empty set, which is the smallest of all possible sets.
- Later we'll see that you can't have an empty group, so the smallest of all possible groups is a group with one object.

We will see that category theory takes this much further.

Family tree

Emphasising relationships in pictures

Family trees are an effective way of making vivid the relationships between people, by drawing lines. Crudely, there are two kinds of line – horizontal ones for brothers and sisters, and vertical ones for parent–child relationships. And perhaps some other kind of symbol for marriages. It becomes more complicated as families become more varied, with remarriages and half-siblings, step-siblings, and so on, not to mention if cousins marry one another.

Drawing a family tree helps explain 'cousin' terminology that is a bit difficult, such as 'second cousin once removed'.

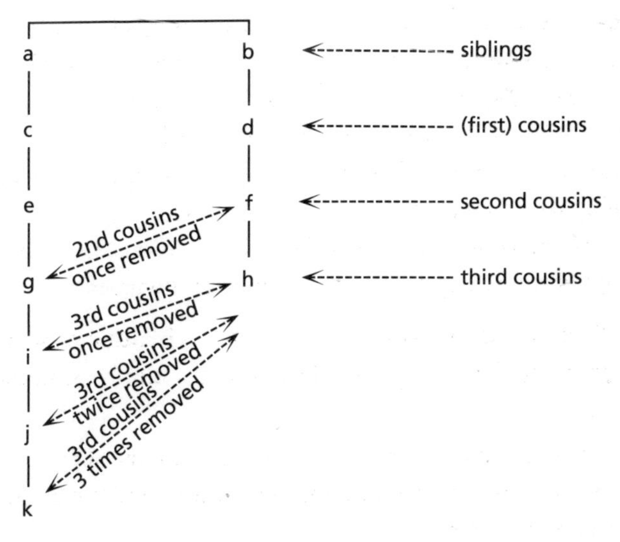

The family tree model can be used in other situations which aren't actually families, but bear some resemblances. My piano teacher had no children, but always said that her pupils were like her children. And in fact, she was such a strong mentor figure, and we her pupils had such a strong shared experience at competitions and masterclasses as well as from her lessons, that we became a bit like brothers and sisters. I think of them as my 'pianistic brothers and sisters' and we have a strong bond that lasts to this day. My piano teacher didn't just teach music, but instilled values and principles in us like parents do (or at least, as they should), and my piano siblings and I will always have that in common. Even when I meet her pupils who are much older or younger than me, so we were never actually pupils at the same time, I feel a bond with them.

Piano family trees are a bit more skewed than real family trees, as people are likely to have either no piano pupils (if they don't become piano teachers themselves) or a very large number of pupils (if they do). This is in contrast to having children, where a large proportion of people have a small number of children. Still, it is fun to trace my ancestry: my pianistic great-grandmother was Clara Schumann, wife of Robert Schumann the composer. This is actually further back than I know my genetic ancestry: I have no idea who my great-grandparents were.

Mathematical family trees are quite a well-known phenomenon, at least among mathematicians. In fact there's a website that tries to trace the mathematical genealogy of all mathematicians in the world, and can generate family trees on request. In maths you count as being 'born' when you get your PhD, and your 'parent' is your PhD supervisor. As with my piano teacher, this resembles a family relationship as well. Many supervisors, at least, good ones (like mine) are very strong mentor figures, who not only guide a student towards a thesis, but shape the way they think and behave, at least intellectually. When I meet my new mathematical siblings, I always feel a connection to them, as with my pianistic siblings; perhaps this is a bit like meeting long-lost brothers and sisters.

Anyway it turns out I can trace my mathematical ancestry back further than my genetic ancestry as well: my mathematical great-grandfather was Alan Turing, the great codebreaker of the Second World War who was treated so abominably afterwards – he was prosecuted for being homosexual, and only granted a posthumous pardon in 2013.

Category theory also represents relationships by diagrams, a bit like family trees, flight maps, street maps and our 'lattice' diagram of factors of 30. The representation is a little simplistic, but that can often seem the case with abstraction – some crucial details have been thrown away. As usual the result is to highlight some feature that we're interested in, in this case particular types of relationships, and to be able to compare those features with other situations.

Category theory represents relationships by drawing arrows, to bring out the structural features of the situation. The arrows represent the relationships in the world that we're currently thinking about, and we can have multiple arrows to represent multiple relationships between the same two things. One of the most powerful aspects of this approach is that it makes everything geometrical, which means we can employ another useful part of our brain to help us do the reasoning.

In fact, when we read diagrams like family trees, we're reading them more *topologically* than geometrically. It doesn't really matter to us what shape the arrows are, it just matters where they start and where they end. Just like if you're taking the Tube, it doesn't really matter where underground the tunnel goes, as long as you can get on and off at the stations you want.

It's remarkable how much insight this approach provides. We are going to draw more and more types of diagrams that look less and less like family trees. Here are some typical diagrams of relationships in category theory. We'll look more closely at what they mean a bit later.

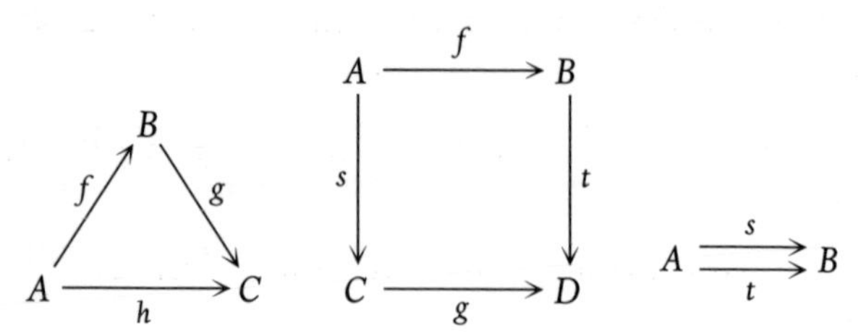

Friends

Relationships that can go both ways or not

You could draw a picture of your network of friends too. You could start by drawing a dot on the page for each friend, and then you could draw a line connecting them up if they're friends. You might immediately run into some curious questions:

1. Is everyone a friend of themselves (or is everyone their own worst enemy)?
2. If someone is your friend, are you necessarily their friend too?
3. Are all your friends' friends necessarily your friends? (Facebook wants to say yes.)

If you decide the answer to the second question is no, then the lines connecting people up better have arrows on them to distinguish between your being someone's friend, and their being your friend. Like this perhaps:

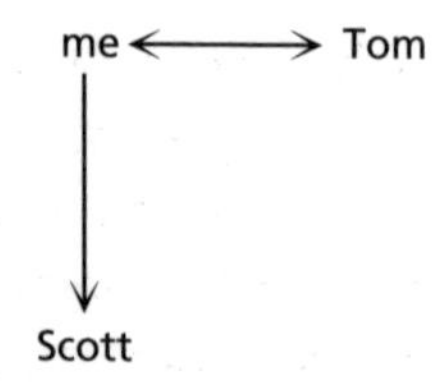

Here I'm Tom's friend, and Tom is my friend. However, I am

Scott's friend, but Scott is not my friend. (Perhaps I am kind to Scott but he is not kind to me.)

Once you've drawn this graph, some features will be very visually clear.

* If you have no friends, you'll just be a single dot on an empty page.
* If you're very popular, you'll have tons of lines emanating from you.

This will be noticeable, because the people joined to you will not have so many lines emanating from them:

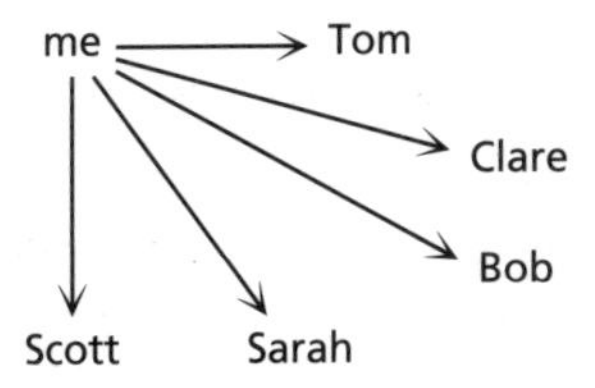

If you're part of a very tightly knit, coherent group of friends, there will be a tight knot of dots with lines going in every direction between them:

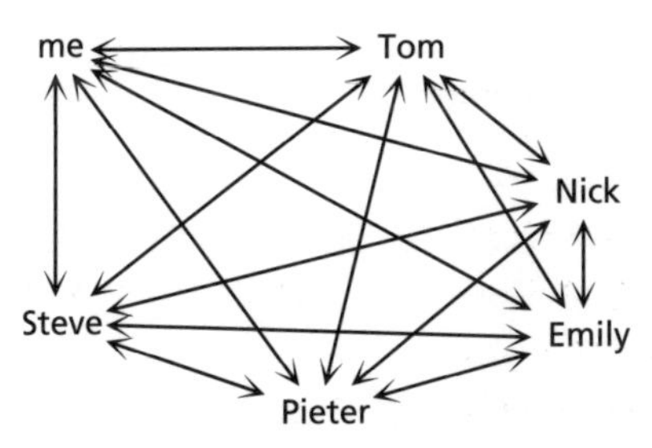

Category theory takes this kind of picture very seriously, but does impose some rules on the types of relationships that it can talk about. They're not exactly the same as the list above, but they're related. The above three questions about friendship charts are the important questions about something called 'equivalence relations' – a particularly important type of relationship. Equivalence relations are very neat and tidy because they always obey three rules; in the case of friendships, that corresponds to answering yes to the three questions above.

The first rule is *reflexivity*, which says that everyone is related to themselves. The second rule is *symmetry*, which says that if A is related to B then B is related to A. The third and last rule is *transitivity*, which we already saw in Chapter 4. This rule says that if A is related to B and B is related to C, then A is related to C.

One example of an equivalence relation is similar triangles. Remember that triangles are similar if they have the same angles but not necessarily the same sides. We can check the three rules:

1. Every triangle has the same angles as itself, so is similar to itself.
2. If triangle A is similar to triangle B, then A has the same angles as B. But then B has the same angles as A, so B is also similar to A.
3. If triangle A is similar to triangle B, that means A and B have the same angles. If triangle B is similar to triangle C, that means B and C have the same angles. But then A and C have the same angles, so triangle A is similar to triangle C.

A more basic (and redundant-sounding) example of an equivalence relation is equality. Again we can check the rules:

1. No matter what objects we're talking about, we always have $A = A$.
2. If $A = B$, then definitely $B = A$.
3. If $A = B$ and $B = C$, then definitely $A = C$.

This is good, because if we're going to think about a broader notion of relation, our basic simpler notion of equality should still be included. It shows that equivalence relations are a *generalisation* of equality. We'll see that the relations allowed in category theory are even more general than this. This is because many relationships between mathematical objects are not as neat and tidy as equivalence relations, but we still want to study them.

Tidying up, or not

Knowing when to leave things in their natural place

When my desk is messy, all the objects are in their natural positions where they have been left. That's what I like to think anyway – I have allowed my sea of papers and copious pens and pencils (I must have at least a hundred of them on my desk) to fill up space in the way they feel most comfortable. However, sometimes I have to tidy up, usually because my 'desk' is actually my dining table and so if I have friends round for dinner I have to clear it up. In that case I try to put the papers into piles, or one big pile. Once they're in that pile, they're neat and tidy and much easier to carry around, but I've destroyed their natural geometry. I'll find it much harder to locate the things I need from that pile, because it's all just lined up in a column, whereas when it was spread out all over my desk, I had a sense of where everything was.

This is one of the important aspects of bringing out the natural geometry of mathematics, as category theory does. It turns an *abstract* notion of 'relationship' into a *visible* notion, an arrow that we draw in a 'map' or other physical representation of the abstract situation. Moreover, it also builds the visible representation up in a way that has shape.

Algebra as most of us know it consists of writing symbols in a straight line, and then in another straight line, and then in another straight line, like several very orderly piles of paper:

$$
\begin{aligned}
2x + 3 &= 7 \\
2x &= 7 - 3 \\
2x &= 4 \\
x &= \frac{4}{2} = 2.
\end{aligned}
$$

However, when we're dealing with more subtle relationships between things, those things don't want to be tidied up into straight lines – they have a natural geometry on the page, and one of the prominent features of category theory is that the natural geometry is allowed to remain.

Here's an example of a piece of algebra-in-a-straight-line:

$$xC.By.zA = Az.yB.Cx$$

It looks rather obscure, but has natural geometry on the faces of a cube. Here the small letters are the faces of the cube, marked with double arrows, the capital letters are the edges, marked with single arrows. The whole thing has a very precise meaning in category theory, which is a bit too complicated to go into just yet, but perhaps you can work out what the rules are for going from the algebra in a line to the diagram with cubes.

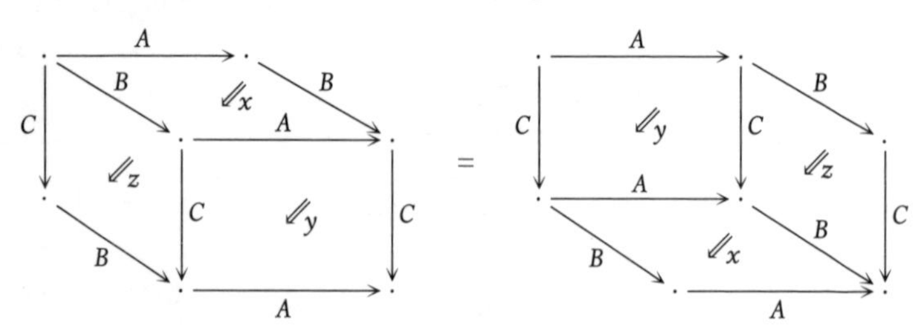

This is really just like directions to build a structure out of smaller pieces – rectangular faces and long, thin edge pieces. If you have rectangular pieces labelled *x*, *y*, *z*, there might be all sorts of different ways to fit them together. But once you've stuck a *C* edge piece onto the corner of the *x* piece (which you might call *xC*), and a *B* edge piece onto the corner of the *y* piece (which you might call *By*), then there's only one way of sticking those two composite pieces together. Likewise the *zA* piece. And so on.

In fact, this has even more natural geometry as pieces of 'string':

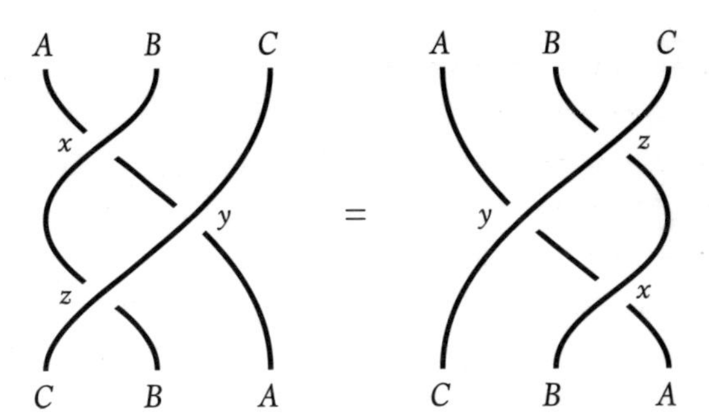

It would be even harder to explain how the string corresponds to the cubes, but perhaps you can see that the left-hand string picture is 'the same' as the right-hand one, in the sense that if it were really made of string that was pinned down at both ends, you could wiggle the string around to get from the left-hand side to the right-hand side. These sorts of pictures are called 'braids' in maths because they're like braids in hair, and some mathematical arguments all boil down to the question of whether two braids are the same in this sense of wiggling string around.

Both the cube diagram and the one with strings are typical calculations in *higher-dimensional* category theory. Even advanced category theorists disagree about which type of picture is most illuminating.

All these sorts of diagrams are a key feature of category theory, especially the ones with arrows. If you even drew just a square out of arrows, like this:

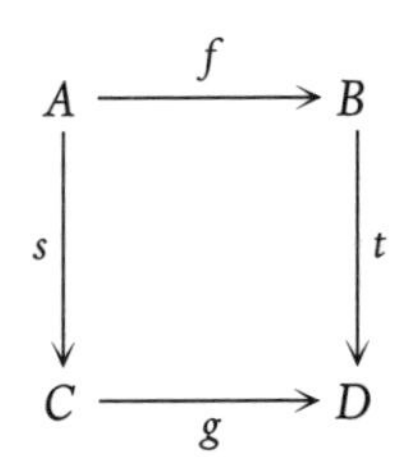

any random pure mathematician would be likely to recognise it as something from category theory.

There's a theory that there are three aspects of mathematics: algebra, geometry and logic. Algebra is, broadly speaking, where we manipulate symbols. Geometry deals with shape and position. Logic deals with making arguments about things. The theory goes that all mathematicians are located somewhere on an edge of this triangle:

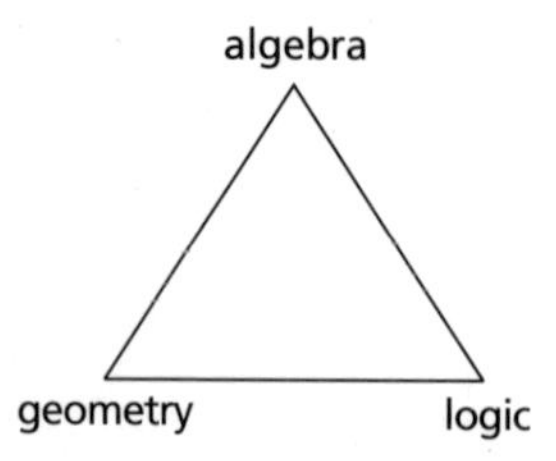

But category theory seems to combine all three of those things. It's about the structure of arguments, and it deals with algebra geometrically.

One-way streets

Showing different types of route in one map

A street map is, in a way, a diagram of relationships between places. The 'relationships' in this case are 'ways of getting from *A* to *B*'. If the map is very detailed, it will have the one-way system marked on it as well, in which case a way of getting from *A* to *B* won't necessarily be reversible to give a way of getting from *B* to *A*.

If that map is *really* detailed, it will also show cycle lanes. Sometimes a route from *A* to *B* will be for both cars and bicycles, and sometimes only for one or the other. It might also indicate bus routes, tram lines and pedestrian streets. Pedestrian streets, of course, are unlikely to be directional. (Although there are places in the Tube where the passage-ways become one-way, such as the connection between the

Central line and the Northern line at Bank, where there are entirely separate spiral staircases for going down and for going up.)

All this is building up a picture of a city less in terms of where things are and more in terms of how you can get from any one thing to any other thing. It's emphasising the relationships between things rather than just the things, like in category theory. One important aspect of this is that things can be related in more than one way. Also, the relationship is not necessarily *symmetric*, that is a route from A to B is not necessarily reversible, because of things like one-way streets. This makes it a different kind of relationship from the equivalence relations that we mentioned above. With relations in category theory there's still reflexivity (things related to themselves) and transitivity, but there's no longer the requirement of symmetry, and there's a new possibility, which is that things can be related in several different ways.

The following diagram depicts a very small category:

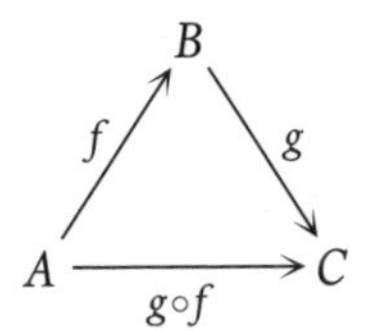

Here f is like a route from A to B, and g is like a route from B to C. Then $g \circ f$ is a shorthand we use for the route from A to C that consists of going along f first and then g. (There are technical reasons we put the f on the right of the g, which I won't go into.)

This is like a train route map from London to Sheffield via Doncaster, except to make it a bit more *geographically* accurate it would look more like this:

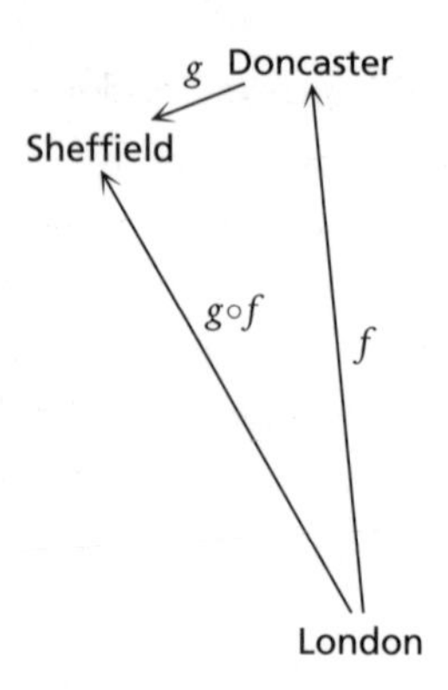

although the precise layout doesn't make any difference *mathematically*. The diagram tells us there's a train from London to Doncaster and a train from Doncaster to Sheffield. You can take one train followed by the next to get from London to Sheffield. These arrows don't represent the physical route that the train takes, but the *abstract* fact that there is a route from London to Sheffield.

In this next diagram there is an extra arrow marked from London to Sheffield, because there is in fact a direct train from London to Sheffield where you don't have to change in Doncaster.

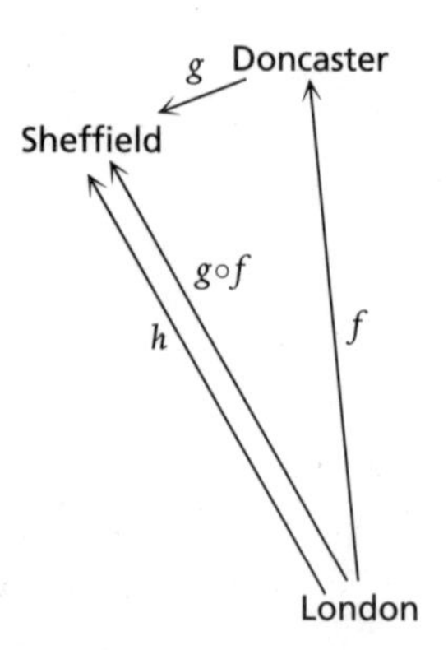

This now shows that there are two routes from London to Sheffield. One of them is the 'compiled' journey involving two trains, and one of them isn't. In category theory, as in mathematics at large, this process of doing one thing and then another is called *composition*.

You might notice that there are some routes on this map that we haven't drawn. For example, you can get from London to London by doing nothing. It's like the reflexivity of relations. You can also get from Sheffield to Sheffield, and Doncaster to Doncaster. We could draw these in as mini-arrows,

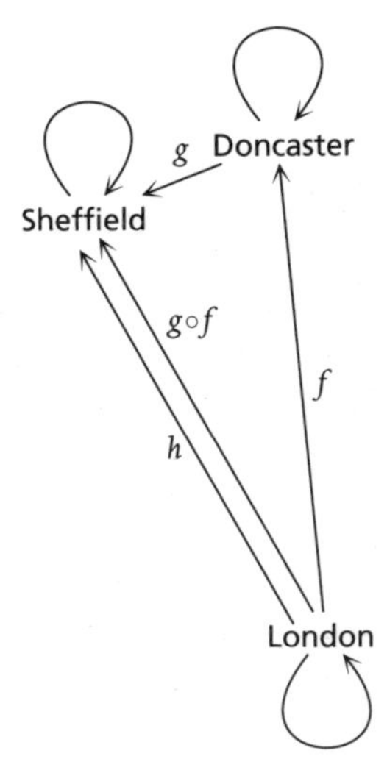

but this would be a bit pointless because they're so obvious.

We will formalise these ideas about relations and drawing arrows at the end of the chapter.

Axiomatisation of categories

Just as we did with groups in Chapter 8, we define categories by axiomatising them. We need to know what the basic building blocks are, and how we're allowed to stick them together.

A *category* in mathematics starts with a set of objects, and a set of relationships between them. Now, these relationships are not necessarily symmetric, so we need to change our wording a bit to bring this out. So, instead of saying a 'relationship between A and B', it would be better to say a 'relationship from A to B' to emphasise that it only goes one way. In fact, in category theory we sometimes say 'arrow from A to B' to emphasise that direction even more, and to remind ourselves of the fact that we draw helpful pictures of these relationships using arrows. We might also say 'morphism' because sometimes these things are more like a way of morphing something into something else, like morphing a doughnut into a coffee cup.

Now we state the rules that our relationships must obey:

1. (A bit like transitivity.) Given an arrow $A \xrightarrow{f} B$ and an arrow $B \xrightarrow{g} C$, this has to result in a *composite* arrow $A \xrightarrow{g \circ f} B$.
2. (A bit like reflexivity.) Given any object A, there has to be an 'identity' arrow $A \xrightarrow{I} A$, so that, for any other arrow, $f \circ I = f$ and $I \circ f = f$.
3. Given three arrows $A \xrightarrow{f} B$, $B \xrightarrow{g} C$, $C \xrightarrow{h} D$, we can make composites in various ways, and they all have to obey the rule

$$(h \circ g) \circ f = h \circ (g \circ f).$$

These rules might also remind you of the axioms for a group, where we also had an identity that 'does nothing', and a rule about putting three things together. What's happened here is that the things we're putting together now are no longer the objects, but the relationships between them. This is a sign that a further level of abstraction or generalisation has taken place – everything has shifted by a level. Level shifting is something that happens a lot in category theory, as we'll see in the chapter on dimensions. It is one of the things that can make you feel like your brain is imploding, or exploding, or getting into some weird contortion like a Möbius strip. And in fact, sometimes mathematicians refer to it as 'yoga'. In the next chapter we'll consider the sorts of ideas that we look at differently once we've turned our brain inside out like this.

Some examples of categories

Here are some small examples of categories. There is a tiny one with just one object and one arrow. Since there is only one arrow, we know it simply has to be the identity arrow. We can draw a picture of this little category:

It doesn't matter very much whether we call the single object x or y or Fred or something else – the picture will still look the same. You might think this is the tiniest and most stupid possible category, but there's an even smaller one that has *no* objects and no arrows, so we can't really draw it. We can't have a category with one object and no arrows, because, according to rule 2 above, any object has to have an identity arrow.

There's a category with only one arrow that *isn't* the identity, which we might draw like this:

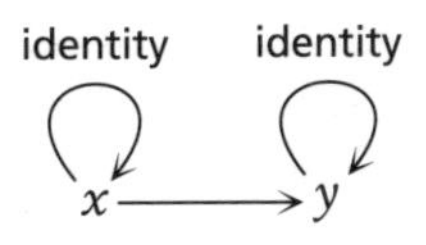

You might realise that it's a bit pointless drawing the identity arrows in all the time, because they're *always* there. So usually we don't bother drawing them because it just takes up space. So we draw the above category like this:

$$x \longrightarrow y$$

Here x and y could be sets and the morphism a function. Or x and y could be groups and the morphism a function that interacts well with the group operation. Or x and y could be topological spaces and the morphism a way of morphing one into the other. However, just as when we turn everything into

x's and y's in algebra, these aren't specific sets or groups or spaces. In an equation x is a *potential* number, and in category theory x is a *potential* set or group or space, or something else, which is why it's just called an object.

Now here's a category in which some arrows can be *composed*:

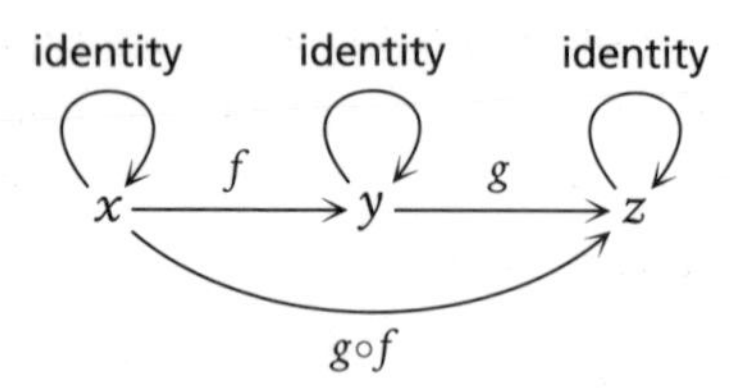

However, not only do we not bother drawing the identities, but we don't really need to bother drawing the *composite* arrow $g \circ f$ either, because we know it has to be there. This is all about making our diagram more efficient, less cluttered and easier to read. So we draw this category like this:

$$x \xrightarrow{f} y \xrightarrow{g} z$$

We'll see in a minute that this 'decluttering' is just like making our lattice picture less cluttered when we were thinking about the factors of 30.

A category with only one object

We can now try to understand that last leap of abstraction that I described having trouble with in Chapter 2. It was about 'one-object categories'. If a category only has one object, then all its arrows start and end in the same place although they're not necessarily the identity:

For example, x could be the set of all integers. There are many different possible functions on the integers that aren't the identity, for example the function that adds 1 to everything, or the function that multiplies everything by 10.

Now, in a category with only one object, *any* arrows are composable, because the end of every arrow matches the beginning of every arrow. So the single object gives us no information and we might as well forget about it. The set of arrows is just a set of things that can be multiplied together but not necessarily divided, like the natural numbers. This is called a *monoid* and so we arrive at the fact that 'a one-object category is the same thing as a monoid'.

Some categories of numbers

We can make a category where the objects are all the natural numbers, and where we have an arrow $a \longrightarrow b$ whenever $a \leqslant b$. So we'll have arrows like these

$$1 \longrightarrow 2 \longrightarrow 3$$

and the composite of these arrows

$$1 \longrightarrow 3.$$

This is a special kind of category in which, given any two objects, there is exactly one arrow between them – because if you think about any two natural numbers a and b, either $a \leqslant b$ or $b \leqslant a$. These can only both be true if $a = b$, in which case we have the identity arrow going from a to itself. This category can be represented as

$$1 \longrightarrow 2 \longrightarrow 3 \longrightarrow 4 \longrightarrow 5 \longrightarrow 6 \longrightarrow \cdots$$

using the earlier principle that we don't need to draw composite arrows or identities. We see that all the numbers end up in a line, just as we expect them to. A category like this is called a 'totally ordered set' because the objects are all in order. Can you see why we couldn't have used $<$ instead of $\leqslant$ for the arrows? It's because we wouldn't have *identities*. There has to be an arrow going from everything to itself, but we don't have

$1 < 1$, so it wouldn't work. In fact we don't have $n < n$ for any number at all.

A different category of numbers comes from the factors of 30 that we looked at before. We could draw an arrow a whenever *a is a factor of b*. In that case we get this picture:

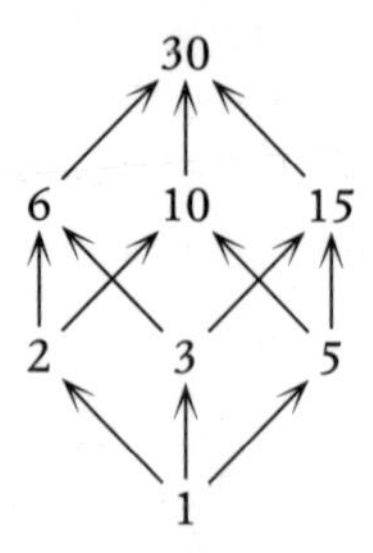

If we try to draw in the composites as well, we get that very much messier picture we saw earlier:

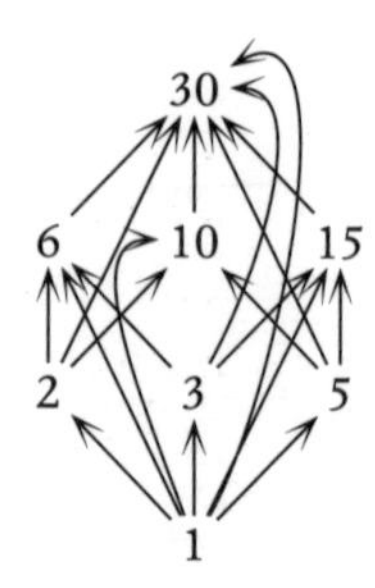

So we see that using a little bit of a category theoretic approach has enabled us to see the structure here more clearly, as it cleaned up our picture. This is one of the fundamental aims of category theory – to 'clean up' our thinking in order to isolate crucial structure. It is quite breathtaking how this framework of objects and arrows opens up endless possibilities, and embraces structures that we might otherwise never have thought of studying in the same light. Here are some examples to show the range of things that can all be encapsulated by this innocuous little picture of two objects and one morphism between them:

$$x \longrightarrow y$$

- Two numbers, and an inequality ($<$ or $>$ or $\leqslant$ or $\geqslant$).
- Two numbers, one divisible by the other.
- Two sets, and a function from one to the other.
- Two sets, one of which is completely contained in the other.
- Two groups, and a function from one to the other that interacts sensibly with the group structure.
- Two spaces, and a way of morphing one into the other.
- Two points in space, and a path from one to the other.
- Two lines in space, and a surface connecting them up.
- A pair of numbers on the left, and the process of forgetting one of them to leave only one number on the right.
- Two logical statements, and a proof that one follows logically from the other.

It might seem that nothing much has been achieved by representing all these situations by this simple picture. However, this is just the starting point in category theory. The next thing we can do is build the pictures up and see what sorts of shapes emerge from multiple arrows and interactions. This is the subject of the next chapter.

12 STRUCTURE

Baked Alaska

Ingredients

1 flat 8-inch round sponge cake

200 g raspberries

1 pint vanilla ice cream

4 egg whites

175 g caster sugar

Method

1. Whisk the egg whites and sugar until very stiff to make the meringue topping.
2. Put the cake on an ovenproof dish and pile the raspberries on top, leaving plenty of space around the edge. Then pack the ice cream on top of that in a dome shape, still leaving a good space at the edge of the cake.
3. Pile the stiff egg whites over the ice cream, making sure there are no gaps, and that the egg whites make a good seal around the cake and all the way down to the dish.
4. Bake in a hot oven (220°C) until the meringue has browned. Eat immediately.

Baked Alaska is not just food – it's science. The various parts of it are not just there for taste, they serve a *structural* purpose. The meringue and the sponge insulate the ice cream from the heat of the oven, so that we get the exciting sensation of eating hot meringue and cold ice cream at the same time.

There are plenty of other types of food that have important

structural features. Sandwiches and sushi, devised to be conveniently edible on the go. Yorkshire puddings the Yorkshire way, where the pudding is essentially an edible plate containing your food. Vol-au-vent, another type of edible food container. Battered fish, where the batter protects the fish from being overcooked on the outside. Or that amazing way of baking a cake on a campfire, inside a hollowed-out orange skin. Not only does the skin hold the cake mix and protect the cake from the fire, but it also gives the cake a lovely subtle orange flavour.

All these are examples of food where the structure is integral to the food, and in some cases where the taste of the food is affected or even determined by the structure. This is different from a cake in the shape of a dinosaur, where the shape is more or less independent of the taste.

One important aspect of category theory is that it examines what part of a mathematical idea is *structural* – more like a baked alaska than a dinosaur cake. It looks very carefully at what role everything is playing in holding the structure together.

Multistorey car park

What the structural part of a building looks like

I was looking at a half-built building with some friends. Actually it was probably even less than half-built – it was just a shell of a structure. We were speculating about what sort of building it was going to be. Some of us were trying to work it out by remembering what we'd read recently about new buildings in the area. But being a mathematician (and a pure one at that) I was staring at it and trying to work it out from 'first principles', that is: what does the thing in front of me actually look like?

I suddenly realised two things. First, it looked like a multistorey car park. Secondly, that *every* building must look like a multistorey car park at that stage in the building process. Usually when I think about the basic structure of a building I think about stripping things away: first the furniture and decorations such as wallpaper and pictures, then windows and doors, then any walls that aren't bearing any load.

But there's the opposite way of thinking about the structure of a building: building it up, rather than stripping it down. Because the structure has to be put in place before any of the decorations go on.

A lot of maths is about structures, and category theory is particularly about structures. What is holding something up? Which parts could you remove, without making the whole thing fall down?

This is a bit like the tale of the parallel postulate, where mathematicians spent hundreds of years trying to work out whether that fifth axiom was actually necessary or not. Would geometry fall apart without it, or would geometry be just the same? In category theory we like to understand exactly what part of the axioms is making everything work in any given mathematical world. This is important as it helps us *generalise* the situation and take it to a slightly different world, if we know exactly what is holding it up.

Here's a thought experiment we can do to see what is holding the integers together. Imagine that the number 2 no longer exists. Which numbers are *now* prime? Remember a prime number is one that is only divisible by 1 and itself, and 1 doesn't count as prime.

Now 3 is still prime, as it's only divisible by 1 and itself. But what about 4? 4 used to be divisible by 2 as well, but 2 *no longer exists*. So 4 is now only divisible by 1 and 4, so it has become 'prime'.

And 5 is still prime, and you might be able to *generalise* this fact now, and realise that any number that used to be prime will still be prime, because it can't suddenly become divisible by new things – there aren't any new things here. (We removed the number 2, but we didn't invent any new things in its place.) The problem will be with *even* numbers because they are no longer going to be divisible by 2 – because 2 doesn't exist.

So 6 is now prime, because it's no longer divisible by 3. Here we have to be a bit more careful about what 'divisible' by 3 means: it means that $6 = 3 \times k$ where k is any whole number. But 6 is no longer 3 times anything, because 2 doesn't exist. So 6 is only divisible by 1 and itself. Likewise 8 and 10.

We now have a curious fact – numbers can now be expressed as a product of 'primes' in different ways. Can you think of an example? Here's one:

$$24 = 3 \times 8 = 4 \times 6.$$

In our new 2-less world, 3, 8, 4 and 6 are all 'prime'. So by throwing away the number 2 we have destroyed one of the fundamental principles of numbers, that every number can be expressed as a product of prime numbers in a *unique* way.

St Paul's Cathedral

Three versions of one structure

I quite often watch television with no sound at the gym, as I prefer listening to music to make me work out harder, but the screens are in front of my face so I can't help watching. One time I watched a really naff docu-drama about the building of St Paul's Cathedral, with stilted auto-subtitles in that crude typeface that brings to mind a robot voice.

I didn't know much about St Paul's at the time except that it was designed by Sir Christopher Wren, and I particularly didn't know how the dome was constructed or how long it took to build or how nearly it didn't get finished. I'm not even sure I appreciated its great and majestic beauty at the time; I just knew it was large and famous.

What I learnt from all this was that the dome is actually made of *three* domes: an inner dome and an outer dome, both visible, and a third, hidden, dome in between them that's actually supporting the structure. The outer dome is the one that's visible across London, still proudly dominating the skyline after all these years despite the arrival of the Shard, the Gherkin, and other, taller buildings. It is not the sheer height of the dome that makes it imposing – it was surpassed as the tallest building in London in 1962, and the Shard is almost three times as tall. The dome is imposing because of its overall size, presenting a severe engineering problem at the time: how to hold such a thing up without the base collapsing?

The inner dome serves the aesthetic desires of the inside of the cathedral, that is, the interior of the cathedral needs a certain balance in its proportions, without a ridiculously huge dome overpowering the main body of space. Until I saw this docu-drama, I didn't realise that the dome visible on the inside was not the same as the one seen from the outside.

But the genius of the construction is the third, hidden, structural dome that 'mediates' in between them. The other two domes are much too broad and flat to be able to support the heavy structure of the lantern at the top of the dome, so in between them is a much pointier brick construction, which would not be very beautiful to look at, but which is strong and secure enough to support the necessary load.

I was a PhD student at the time, and I had this epiphany that this was exactly like the thesis that I was in the process of writing. My thesis involved three expressions of the same structure, one with 'internal' motivation (the internal logic of the situation), one with 'external' motivation (the applications), and the other which was 'hidden' and whose only purpose was structural, to mediate between the two.

The personal part of this drama was that, apparently, Wren had no idea how he was going to achieve the effect he wanted. The building of the cathedral was actually under way, and Wren still had no idea how he was going to construct the dome; he just had a vision of what he was going to accomplish. The idea of three domes came later.

I now have a strong belief in the difference between internal and external motivation, structural mediation between the two, and the idea that if one has a genuinely good idea, the means of accomplishing or justifying it will come later. And that one can be on the point of spectacular failure just before spectacular success. And that I love St Paul's Cathedral.

Category theory often studies different aspects of the same structure. It can be fascinating to turn things inside out and see them from the other way up – understanding something from only one point of view is far too restrictive. The greatest leaps forward in the history of mathematics have often been when connections are made between apparently unrelated subjects, enabling communication and the transfer of both information and techniques. It's like the difference between building a bridge between two islands, and a bridge to nowhere.

Category theory grew up from the study of *algebraic topology*. We have already met various ideas from topology, including surfaces, knots, bagels, doughnuts and the idea of 'morphing' shapes into other shapes as if they were made of playdough. We've also met various ideas from algebra, including groups, relations, associativity, and so on.

Algebraic topology is like a road between those two 'cities', algebra and topology. The original aim was to use algebra to study topology, but then it turned out to be a two-way road, so topology can also be used to study algebra. Category theory helps translate between the two cities. It enables us to ask questions like:

* Are there features in one city that resemble features in another city?
* If we take our tools and techniques from one city to the other, will they still work?
* Are the relationships between things in one city at all like relationships between things in the other city?

Category theory doesn't necessarily answer all those questions, but it gives us a way of posing the questions, and helps us see which ideas are important and which are irrelevant to finding the answers.

CD

Which part makes it a CD?

I once decided to try and remove the label from a CD. I can't remember why – perhaps it was so ugly I couldn't bear to look at it any more? I had been making my own CDs for the first time and so had a pack of self-adhesive CD labels that I really liked using. I think my plan was to design a new label and stick it on the CD myself. I tried sticking a new label on top, but I could still see the old one underneath.

If you think this sounds like a made-up story, I sympathise; I feel a bit like I'm making this part up myself. The thing is, I now can't for the life of me remember why I was trying to remove the label from the CD, but I definitely remember what happened next. I took the label off, and all I was left with was: a transparent piece of plastic.

I felt very foolish. Was it obvious to everyone on earth except me that the crucial part of a CD, the shiny part, was actually structurally part of the label? That apart from this the CD was just a piece of plain plastic?

Similar things have happened to me with dresses, when I've thought 'That is a great dress apart from that ugly flower attachment on it.' But when I investigate the possibility of simply removing the flower, I discover that it's so deeply attached to the dress it's actually part of its structure. The dress stays in the shop.

In category theory one of the important aspects of looking at structure is to see what will go wrong if you discard parts of the structure. This is all part of finding out exactly how something works in case you find yourself in a (mathematical) world with less structure. It's a bit like learning how to whisk egg whites by hand as well as doing it using an electric whisk. It means you'll be able to do it even when you're in a kitchen with no electric whisk. Or no electricity. Perhaps you're in the forest and you really need stiff egg whites? Oh never mind.

One mathematical version of the 'electric whisk' is related to how we solve quadratic equations. We saw in Chapter 7 that we can try to solve the quadratic equation

$$x^2 - 3x + 2 = 0$$

by recognising that the left-hand side can be factorised:

$$x^2 - 3x + 2 = (x - 1)(x - 2).$$

Then we conclude that one of the two brackets must be 0, in order for the answer to be 0, so either $x - 1 = 0$, in which case $x = 1$, or $x - 2 = 0$, in which case $x = 2$. So these are the two solutions.

However, suppose we were doing this on a six-hour clock. You can try putting in some other values for x to see what the answer is. For example, if you put $x = 4$, you'll get

$$\begin{aligned} x^2 - 3x + 2 &= (4 \times 4) - (3 \times 4) + 2 \\ &= 16 - 12 + 2 \\ &= 6 \end{aligned}$$

but on the six-hour clock, 6 *is the same as* 0, so $x = 4$ actually gives 0 as the answer here. You can check that $x = 5$ gives 12, which is also the same as 0. This means that $1, 2, 4$ and 5 are *all* solutions to this quadratic equation on the six-hour clock. What is going on? Where are these 'extra' solutions coming from? How can we look for them and how can we be sure we've found them all?

The key is to go back and carefully look at how this argument works. The crucial moment is where we declare that 'one of the brackets has to equal 0'. What we're saying there is that if we multiply two things together and get 0, one of them had to be 0 already. However, while this is true with normal numbers, it is *not* true on the six-hour clock. For example,

$$3 \times 2 = 6 = 0$$
$$4 \times 3 = 12 = 0.$$

This is why some new solutions have popped up even though when $x = 4$ neither of those brackets $(x - 1)$ and $(x - 2)$ is 0. The point is that when $x = 4$ the brackets work out to be 3 and 2, and when $x = 5$ they work out to be 4 and 3. So those two 'extra' ways of multiplying numbers to get 0 give two 'extra' solutions to the quadratic equation.

We have gone to a mathematical world without a piece of structure that we're rather used to: the fact that the only way to multiply numbers to get 0 is if one of the numbers we're multiplying was already 0. So we have to be careful how we proceed in this other world, and also in *any* other world that doesn't have this structure. We have isolated an important piece of structure that it's important to look for if we want to go round solving quadratic equations in different worlds. Although there might be more solutions floating around for us find, we have to work a lot harder to make sure we've found all the right ones if we don't have this rather useful piece of equipment.

Money

Being careful how you spend it

If you have a lot of money – I mean, really a lot of money – you don't ever have to find out how anything works. If it goes wrong, you can just throw money at it to get it fixed. You can either pay someone else to fix it, or you can just go right ahead and buy a new one. If you're rich, you also don't have to worry about exactly how much money you're spending on things every day, although some rich people apparently still do.

But if you're a normal person, you do have to worry about these things, at least if you want to avoid financial catastrophe. Even if you're not extremely frugal all the time, it's good to be aware of what you're spending money on, so that you can rein it in if necessary.

Some mathematics is done in the 'rich' way – with no fear of ever running out of (mathematical) resources, so without really paying attention to which resources are being used. In contrast, category theory is like being frugal or at least aware of your mathematical spending. That is, the aim is to study mathematics always being aware of what structures you are using to get by at any given moment. You might not be using them explicitly, but sometimes the hidden usage is even more important, precisely because it's hidden, so you're likely to use it without noticing. A bit like when people accidentally get huge credit card bills because their children have bought extras in a game on their mobile phone. Or when you run up a huge roaming bill because your phone has connected to the internet when you're abroad.

Category theory aims to keep track of resources, not because resources might suddenly run out in mathematics (that's not how mathematical resources work, fortunately), but so that you can deliberately go to a planet with fewer resources. The aim is to make connections between different mathematical worlds, and develop techniques that can be used without extra effort in those different worlds.

This is just like the example with quadratic equations that we just saw. The resource in this case is the property

If $a \times b = 0$, *then* $a = 0$ *or* $b = 0$ (*or both*).

Now if you think you will *never* end up in a world without this resource, then you will not care about how many times you use it. But if you care about *modular arithmetic* (on a clock face) or even just the possibility of going to worlds without your resource, then you have to go back through all the techniques you love, and work out when you used this principle, and how to get round it.

A more profound mathematical example involves something called the 'axiom of choice'. This axiom says it is possible to make an infinite number of arbitrary choices. In normal life you might think it's perfectly possible to make an arbitrary choice – it's just like picking a raffle ticket out of a hat. The axiom of choice says it's possible to pick a raffle ticket out of each of an infinite number of hats, which might seem a bit odd to you. Mathematicians don't really agree on whether this is odd or not.

Processes involving some notion of 'infinity' always require great care if we're trying to make them rigorous, and this one about arbitrary choices turns out to be particularly difficult to pin down, which is why it is an axiom all by itself. People are a bit undecided about whether it should be assumed to be true or not, and so the best approach is to *be aware* every time you need to use it.

One branch of category theory deliberately goes to worlds where this axiom is *not* true, to see how much of mathematics can still be done.

Skeleton

Not a whole person, but the last part that remains when all is stripped away

A wonderful old professor sat next to me at dinner in Cambridge one day when he was about ninety. It was round about the time of the scandal at Alder Hey hospital, when it was discovered that, shockingly, organs from dead children had been removed and kept by the hospital without authorisation. The professor told us he was worried that this scandal would put people off organ donation, and that this had moved him to contact Addenbrooke's, the Cambridge teaching hospital, to ask if there was anything at all useful they could do with his old body after he died. He was too old for organ donation, but they told him that his skeleton would be useful for teaching medical students, so he should try not to die mangled in a road accident. (He told us this with typical glee and a twinkle in his eye. I wonder if I will be able to speak with such cheekiness about my own future death.) A few years later I heard that he had passed away at home; I hope that his skeleton is indeed now being used for teaching purposes.

A skeleton is not a whole person, but it's an important part to understand, in order to understand how a person functions. It gives a person their structure. It has little to do with thought, emotions, feelings, and so on, but it's the frame on which everything hangs. This is the point of studying structure in mathematics as well.

Logic is a branch of mathematics that studies the structure of mathematical arguments. Category theory, on the other hand, studies the structure of the mathematical objects themselves. They are similar in a way, in that they're both even more abstract than mathematics itself, as they study the way mathematics is done. However, logic is more obviously used in ordinary daily life – or rather, it is *usable* even if it's often used rather badly. Any time you construct an argument, justify your point of view or make a decision, some element of logic could (or should) come into it, when you start from some more basic thoughts and proceed to some more complex ones.

It is less obvious how the study of mathematical structure could arise in daily life. However, it is the mental exercise of stripping away layers to reveal important structure that is usable everywhere. It also goes the other way round, as we have the mental process of starting with simple structures and carefully building up more complex ones. Category theory formally only does this for mathematical structure, just like formal logic – it only really applies to *mathematical* arguments and not normal arguments in every day life. However, the mental exercise in the abstract mathematical environment prepares us for the concrete non-mathematical environment, just as working out in a gym can make us more generally fit for the world outside the gym.

Battenberg cake

An example of a ubiquitous piece of structure

Here's an example of a mathematical structure that pops up all over the place in different guises. Let's start by thinking about addition on a two-hour clock, or, to use the technical term, 'addition modulo 2'. This means that there are only two numbers, 0 and 1. Now 2 counts as the same as 0, as do 4, 6, 8, 10, . . .; and 3 counts as the same as 1, as do all the odd numbers.

We can now draw an addition table for this. We only need the numbers 0 and 1 (because all other numbers are the same as one of these). And we need to remember that $1 + 1 = 2$ but that 2 is the same as 0, so in fact $1 + 1 = 0$. The addition table then looks like this:

+	0	1
0	0	1
1	1	0

In fact, this is the second smallest possible *group*. We have already seen that the smallest possible group has only one object, the identity. Now we have a group with two objects. This is related to the question we posed at the end of the chapter on principles, about filling in the squares with colours, where each colour can only appear once in each row and each column.

Now here's another way that this pattern appears. We can think about just using the two numbers 1 and -1, and combining them using multiplication. What table does that give us?

×	1	−1
1	1	−1
−1	−1	1

If you compare this with the previous table, you'll see that it has the same pattern, just with different labels in the boxes. We can also think about the rotational symmetry of a rectangle. A rectangle only has two forms of rotational symmetry: the rotation by 0° and the rotation by 180°. If we do the 0° rotation followed by the 180° one, then the result is rotation by a total of 180°. Likewise if we do it in the opposite order. However, if we do a rotation by 180° and then another, we have gone round 360° and we get back to exactly where we started – the same as doing a rotation by 0°, or nothing at all. We can now put these in a table as well:

rotation	0	180
0	0	180
180	180	0

You might not be surprised to see that it's the same table again. We have already seen this pattern in the chapter on context, when we thought about multiplying positive and negative numbers, or real and imaginary numbers, and we drew up the following tables of results:

×	positive	negative
positive	positive	negative
negative	negative	positive

×	real	imaginary
real	real	imaginary
imaginary	imaginary	real

In fact, all the inside part of these tables have the same pattern as a Battenberg cake:

which is designed for the same reasons – we don't want two squares of the same coloured cake to touch each other.

Battenberg challenge

Here's a challenge: can you draw a picture of a Battenberg cake, each of whose mini-cakes is itself a Battenberg cake? I call this the 'iterated Battenberg'. This means you have to start with two types of Battenberg cake, in different colours. So there are four colours altogether and they need to fit together in a 4×4 grid. In fact, we've seen one of these already at the end of Chapter 3. There we had four examples of 4×4 grids of colours, and the first one was an iterated Battenberg.

This pattern comes up if we look at the rotations *and* reflections of a rectangle, instead of just the rotations. Another place this comes up is if we draw a multiplication table for the odd numbers, modulo 8. We only need to consider the numbers 1, 3, 5, 7, as all other odd numbers will be the same as these on the eight-hour clock. You can try filling in this multiplication table, remembering that every time you get to 8 you go back to 0. So 3×3 is 9, which is the same as 1, and so on.

×	1	3	5	7
1				
3				
5				
7				

You should get the following table, with the iterated Battenberg pattern:

×	1	3	5	7
1	1	3	5	7
3	3	1	7	5
5	5	7	1	3
7	7	5	3	1

Now that we have found Battenberg-type structures all over the place, we need to see what it means to say that all of these structures are 'really the same'. One of the easiest ways to see that they were all the same was to isolate the structure and put it in the tables as we did above. Category theory does something similar for more general forms of structure. We have already seen how we draw relationships between objects using arrows. We can now boil a piece of structure down to a little diagram of arrows.

For example, we might go round looking for diagrams like this:

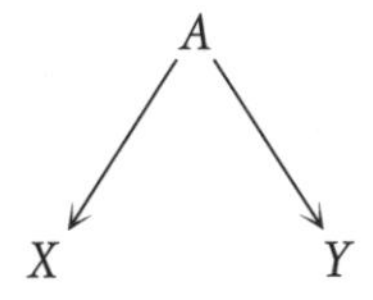

or like this:

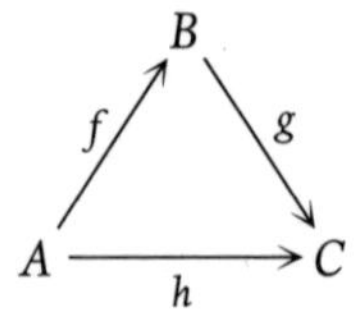

In the latter case, we might use rotation by 180° for f and g, which gives us this diagram:

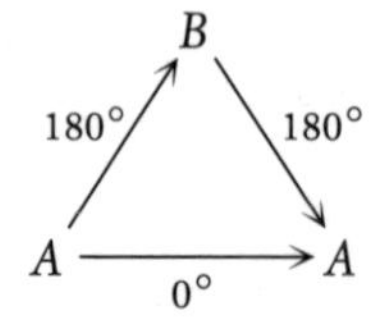

showing that if we do rotation by 180° twice, it's the same as doing nothing.

In the same way as putting all the above examples in 2×2 tables, these category theory diagrams help us to see the structure in different situations, and we can then more easily see if it's 'the same' as some other piece of structure in an otherwise completely different situation. But what does 'the same' mean? This is the subject of the next chapter.

13 SAMENESS

Raw chocolate cookies

Ingredients

100 g dried unsulphured apricots

50 g pitted dates

60 g ground almonds

100 g cornflour

90 g raw cocoa powder

60 g raw cocoa butter

15 g coconut oil

Method

1. Gently melt the cocoa butter and coconut oil.
2. Chuck everything in the food processor and blend until it resembles cookie dough.
3. Press the dough flat onto a sheet of baking paper dusted with cocoa powder, and then roll out until quite thin.
4. Cut into small squares and chill until firm.

We have already seen a recipe for gluten-free chocolate brownies. But what if we also want to make them vegan? Sugar-free? Low fat? All of these things are possible, but the result becomes gradually less and less like actual brownies. Each time you make one substitution the result might be similar, but as you make more and more of them, you get further and further away from the original concept.

The above recipe is not only gluten-free, vegan, sugar-free

and low fat, but also raw. Aside from the arguments about the health benefits of eating raw food, the *taste* benefits of raw chocolate are clear to me – unroasted cocoa is delicate and fragrant, which is why I came up with this recipe. It's not clear that the name 'raw cookie' makes any sense though, as a 'cookie' is something that is 'cooked', according to its name. But these chocolate thingies are similar to cookies in other ways: the texture is similar, the flavour is similar (but better in my opinion) and they play a similar role in my daily diet – a treat, a snack, something to go with my coffee.

One of the key aims of category theory is to make precise, slightly subtle notions of sameness. As I discussed before, 'equality' is a rather stringent notion, and really not very many things are genuinely, rigorously equal to each other – you can only actually be equal to yourself. However, there are things we consider to be more-or-less the same in certain situations.

Category theory highlights the relationships between things, so it enables us to look for more subtle notions of 'sameness' than equality, via these relationships. These versions of equivalence are particularly prominent kinds of relationships between things in category theory. The context we're thinking about is crucial, of course, as some things will be effectively the same in some contexts but not others. One of my favourite examples of this being misunderstood is where computers try and judge things as 'the same' on behalf of human beings, in online shopping.

Online shopping

Better and worse substitutions

Online grocery shopping has revolutionised my life. I'm really terrible at grocery shopping because I get tempted by all the delicious-looking things on the shelves, buy things that will make me put on weight, and spend too much money. When I do my grocery shopping online, however, I'm not at all tempted by the offers that flash up in front of my face (although I might be in trouble when they invent a way of making the smell of freshly baked cookies waft out of my computer screen). Moreover, I don't have to carry my shopping home.

However, the companies' method of substitution is a bit suspect, in my opinion. You know how it works: if something you've ordered is not available, they bring you something else instead, but you're allowed to decline it if you want.

Once, just before Christmas, I ordered four 500 g bags of brussels sprouts. Yes, I eat loads of brussels sprouts – I find them delicious and filling and they're so healthy. For a treat I sometimes dip them in barely sweetened homemade dark chocolate. Anyway, they didn't have any 500 g bags of sprouts, so they substituted four 100 g bags of sprouts instead. Yes, four bags in total, so I had a total of 400 g of sprouts instead of 2 kg.

But the funniest substitution I've heard about was my friend who ordered a toothbrush and was presented with a toilet brush instead. The computer system was thinking too much in terms of the inherent characteristics of the objects ('they're both brushes') instead of the role that they fulfil.

We have already seen that category theory studies things in context, via their relationships with other objects rather than just looking at what an object is like by itself. One of the aims of this is to be able to make precise which things count as 'the same' in particular contexts. This is at the very heart of mathematics. As a basic example, this is really what solving equations is all about. You start with a statement in which something is equal to something else, and you replace it with successive statements in which something is equal to something else in a progressively more useful way, until you have some particularly useful information at the end.

Equations give us different ways of understanding the same concept. For example,

$$3 \times 4 = 4 \times 3$$

tells us that if we take 3 bags with 4 apples in each, we'll get the same number of apples as if we take 4 bags with 3 apples in each, although these two situations are not *exactly* the same. Similarly the equation

$$5 + (5 + 3) = (5 + 5) + 3$$

tells us that if we do $5 + 3$ and then add 5 to that, it's the same as first doing $5 + 5$ and then adding 3 to that. Again, these are not *exactly* the same processes. In fact, that's why this equation is useful – because the second way (doing $5 + 5$ first) is probably easier, as it gives 10, and then you have to do $10 + 3$. If you follow the left-hand side of the equation, you end up having to do $5 + 8$, which is probably more difficult for most people.

So we see that this equals sign is already hiding a lot of information. It doesn't mean that the left-hand side is *exactly* the same as the right-hand side, because it visibly is not. It just means that if you follow the process on the left, you'll get the same answer as if you follow the process on the right. This gives us the slightly uncomfortable fact that the only genuinely honest equations are the ones where the left-hand side is *exactly* the same as the right-hand side, such as

$$1 = 1$$

or

$$x = x$$

and these equations are completely useless. The only useful equations are those that tell us two *different* ways of doing something are 'somehow the same'.

One of the aims of category theory is to make precise different notions of what 'somehow the same' can mean, taking into account that different meanings are useful and relevant in different situations. In category theory we observe that sometimes when we say things are 'equal' we're not being entirely honest. It's a sort of white lie that doesn't matter too much until you get into more delicate situations, and then your white lies start piling up and you need to start keeping track of them. One of the reasons you don't usually study category theory until you're an advanced undergraduate or masters student of mathematics is that you can mostly get away with ignoring the piles of mathematical white lies up until that point, without getting into too much trouble.

Here are some examples of notions of 'sameness' we've already seen that are not precisely 'equality':

* Similar triangles, which have the same angles but different lengths of sides.
* Topological sameness, in which a doughnut is 'the same' as a coffee cup because one can be squashed into the shape of the other.
* The symmetries of an equaliteral triangle, and the ways of ordering the numbers 1, 2, 3, because we can label the corners of the triangle with 1, 2, 3 and see where they move to when we flip or rotate the triangle.
* The various different versions of the Battenberg cake that we saw in the last chapter: coming from addition modulo 2, multiplying ± 1, positive and negative numbers in general, real and imaginary numbers and rotations of a rectangle.

Sometimes in category theory the process goes the other way round – instead of asking what counts as 'the same' in a given context, we start by knowing what we *want* to count as the same, and ask what context will make that true. Sometimes it's not the most obvious one. For example, the most obvious context (or rather, category) in which to study shapes like doughnuts and coffee cups does *not* result in the doughnut and coffee cup counting as 'the same'. The fact that we want them to count as the same means that mathematicians have built much more subtle categories in which to study them. In fact, the *theory* behind building these more subtle categories is an important piece of maths in its own right, which is one of the main areas of current research in the field.

In this chapter we'll see how category theory makes these ideas precise.

Nelson's message

Sacrificing some sameness for a greater good

At a key moment just as the Battle of Trafalgar was about to get under way on 21 October 1805, Lord Nelson sent out his now famous message to rouse and inspire his fleet:

> *England expects that every man will do his duty.*

This was raised in a flag signal before they sailed to their famous – but, for Nelson, fatal – victory. However, Nelson's original message was

> *England confides that every man will do his duty,*

which does have a slightly different tenor to it. It is worth remembering that 'confide' was used in a sense that has more or less died out now: he did not mean that England was telling a secret; he meant that England was confident that every man would do his duty. This has a different tone to 'expects' – perhaps it is more trusting. It is not a command, not even an implied command, it is a simple statement of confidence in the fleet, quite a British understatement I think. Not 'Go out there and defeat the enemy!' Imagine if someone says to you, before a big event, 'I expect you will be brilliant', as opposed to, 'I am confident that you will be brilliant'.

Anyway Nelson asked his signal lieutenant, John Pasco, to relay this message to the fleet in flag signals, asking him to be quick as he had one more signal to make afterwards. Pasco respectfully suggested the word change, for the sake of efficiency. The point was that 'expects' was in the signal book and could be signalled in one go, whereas 'confides' would have to be spelt letter by letter – much more arduous and time-consuming. Nelson authorised the change.

The message was equivalent enough for him, in meaning. But to the signal lieutenant, the new message was much simpler.

Often in mathematics the aim of finding things that are more or less the same in a given context is similar: we can then replace an object in our thoughts (or calculations) with one that is more or less the same in the given context, but much easier in some other respect. Perhaps it is simpler to work with or simpler to draw or simpler to think about.

For example, topologically an infinitely large piece of paper is the same as a very small piece of paper. In fact, they're both the same as a single dot. It is very useful to be able to swap between these things in different situations, knowing that *topologically* they're all the same. Sometimes a single dot is the simplest thing to think about because it really is very tiny. But sometimes a whole 'piece of paper' is more useful. In life this is because you can actually draw something on it (unlike on a tiny dot), and in maths it's actually quite similar. Here, by 'piece of paper' I really mean a flat square surface. These are very useful objects in topology because they are building blocks that we can use to make other surfaces like a patchwork. We couldn't do that with dots, because when you stick a dot to another dot you just get a dot, because the second dot has nowhere to go except right on top of the first dot. If we tried to build surface out of dots we'd never get anywhere. This is like trying to build something out of Lego when all you have is the tiny 1-by-1 pieces. All you'll be able to do is stack them up in a narrow tower. With dots it's even worse because they have no height, so you will go neither sideways nor upwards.

Here's how this fact is expressed technically in mathematics. The notion of sameness we're using here is the 'playdough' one, which is called *homotopy equivalence*. The mathematical version of a piece of paper is a plane. So we say that a plane is homotopy equivalent to a point.

Building up spaces by gluing together smaller ones is a process called 'taking colimits'. And the mathematical stumbling block we have here is that 'taking colimits does not preserve homotopy equivalence', which means that although a plane is more or less the same as a dot, you can stick planes together in such a way that is very different from sticking dots together. For example, you glue two pieces of paper along pairs of edges to make a cylinder. A cylinder is very different from a dot, because it has a hole in it.

Chocolate cake

When small differences add up to big ones by mistake

If you offer a small child a choice of several pieces of chocolate cake, they will almost certainly be completely sure which one is the best. If you give them one that wasn't the best, they will be upset and possibly cry.

Now imagine weighing the pieces of chocolate cake. You can imagine that if you offer the child one piece that weighs 100 g and another that weighs 95 g, they might not notice the difference. So those two pieces are 'more or less the same' to the child as well as to you. Now you could offer the child the 95 g one and a 90 g one, and they might still not notice the difference. Then 90 and 85. Then 85 and 80. And so on. You could keep going like this all the way down to 50 g, but then if you showed them the first piece, the 100 g piece, they'd say it was definitely bigger.

What has happened? Something odd has happened that isn't supposed to happen when things are the same. If you have

$$a = b$$
$$b = c$$
$$c = d$$
$$d = e$$
$$\vdots$$

and so on, you can keep going forever, up to, say,

$$y = z$$

and you'll still have $a = z$. Not so with the child's chocolate cake. This is a problem. So category theory doesn't allow any old thing to be a notion of 'sameness'. The chocolate cake one doesn't work, for example. We would have to use a different axiomatisation to encapsulate that situation.

Category theory wants to use notions of sameness that operate enough like equalities that we can manipulate them a bit like we are used to doing with equalities, just with perhaps a little more care. This means we should be able to use chains of sameness, as above, and we should be able to substitute things that are 'the same' and get a result that is 'the same', like when we use potato flour in a brownie recipe and get something that is still more or less the same as a brownie.

In category theory we are able to express these notions of sameness using the 'relationships' between objects that we have. Remember that we draw these as arrows and actually call them arrows or morphisms. Some of the arrows might be not at all like sameness. For example, we have looked at all numbers, and drawn an arrow $a \longrightarrow b$ whenever $a \leqslant b$.

Now obviously some of these arrows aren't like 'sameness' at all, because we have things like $3 \leqslant 10$ but 10 is not at all like 3. Unfortunately this isn't a very interesting example for us to think about, because numbers are so basic that the only notion of sameness in this category is in fact equality. In order for us to think about more interesting notions of sameness we need to think about the relationships between objects more like a process of getting from *A* to *B*, like a route through a city. The question now is:

Is the process reversible?

In category theory, things only count as 'more or less the same' is you can reverse the process of getting from *A* to *B*. If you can only go one way and not get back again, it doesn't count.

Frozen egg

Processes that are nearly reversible

When you melt chocolate carefully enough, you can always let it set it again and it will be pretty much back to how it was when it started. Butter is a bit more tricky – it is likely to separate, and then when it sets again it won't be quite the same.

What about ice cream? You're not supposed to melt and refreeze ice cream in case you get food poisoning, but I've done it plenty of times (not wishing to waste the ice cream) and the re-frozen ice cream seems just the same as before, to me. And it's never made me ill (yet). However, it does lose a bit of air when it's refrozen, so the resulting ice cream is a bit more solid than before.

So much for taking something frozen, thawing it and freezing it again. What about freezing things that aren't supposed to be used frozen, and thawing them again? This works fine with water of course, and you can keep doing it as many times as you want. Milk can look a bit suspect after you thaw it again – if it wasn't homogenised then it separates when it thaws and looks disgusting, like it's gone off. I'm still happy to use it like that for cooking, but wouldn't give it to someone to put in their tea, as they'd probably think I was crazy.

Have you ever frozen an egg? The result after thawing is slightly unnerving. The white seems to go back to looking completely normal, but the yolk does not lie in a flattened little blob as you would expect a raw egg yolk to do. It stands out from this bowl of egg white as if it were a boiled egg yolk. The first time I tried this I cut it in half and it even looked like a boiled egg yolk on the inside. I can't remember what it tasted like but I must have tried it, knowing me. The thing is that I mostly only eat egg whites, so it didn't really matter to me that the yolk had become peculiar. To me the frozen-and-thawed egg was just as good as a normal egg. (Actually it was even better, as it was much easier to remove and discard the yolk in this weird pseudo-boiled state than when it's raw.)

The point about all this is that freezing water is an entirely reversible process, but the other processes are only 'more or less' reversible. That is, when you try and undo the process you get something that is only 'more or less' the same as what you started with. This is something that category theory can deal with, as it deals with good notions of 'more or less the same'. There are plenty of occasions when something gives you not exactly the right answer but more or less the right answer. Category theory gives us a way of saying this precisely without having to wave our arms around a bit and mumble 'Um, sort of... '.

In maths, instead of saying 'reversible', we say 'invertible'. One mathematical process that is invertible is 'adding 2'. We could draw it as a process, like this:

$$3 \xrightarrow{+2} 5$$

and then we can show the reverse process like this:

$$5 \xrightarrow{-2} 3$$

and to show that this really gets us back to where we started we could draw this:

$$3 \xrightarrow{+2} 5 \xrightarrow{-2} 3.$$

Actually in maths we're interested in more than just getting back to where we started – we want to know if the *process* of going there and back is the same as the *process* of never going anywhere in the first place. This doesn't make a lot of sense with numbers, because our processes aren't subtle enough to pick up that kind of difference. It's the kind of delicate situation that only really comes up when you study things more delicate than numbers.

Still, here's something that *isn't* invertible. We can think about squaring numbers. For example,

$$3 \xrightarrow{\text{squaring}} 9$$

but we could also have

$$-3 \xrightarrow{\text{squaring}} 9.$$

So when we reverse this process starting at 9, how do we know whether the answer should be 3 or −3? We don't. So squaring is not *invertible*.

Custard

When combining things in a different order makes a difference

Some recipes require you to separate the egg yolks from the egg whites. Sometimes this is because you're only using the whites, like in meringue, or only the yolks, like in custard. Sometimes you're using both, but separately, in a pleasing sort of coherence, like with lemon meringue pie, where you use the yolks in the filling and the whites in the meringue topping. Other recipes require you to separate them just so you can mix them together in different ways, like with chocolate mousse, where the yolks get mixed with the chocolate, and the whites are whisked to stiff peaks and folded in.

When you're making custard (and many other things with separated eggs) you absolutely have to do everything in the right order and the right combinations. You start by whisking the egg yolks with the sugar, and then whisking in the milk. If you started by whisking the sugar with the milk, and then whisking in the egg yolks, it wouldn't be the same at all.

Making cake is much less sensitive. I usually start by creaming the sugar and the butter, then adding the eggs and then the flour. But actually you could start by whisking the sugar and the eggs, and then adding the butter, although it won't blend as well unless it's melted. In fact with the advent of electric whisks and food processors all these techniques are fairly unnecessary – you can basically just chuck everything in the food processor at once and press go.

We could represent the making of custard by a diagram like this:

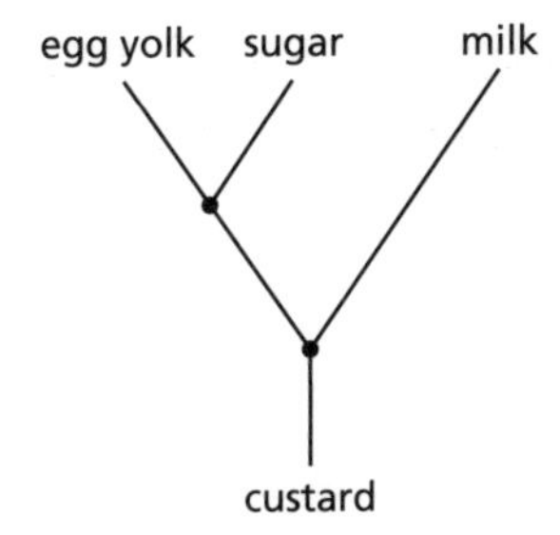

and we could then observe that if the sugar 'branch' were attached to the milk 'branch' first instead of the egg yolk branch, it would not be the same. That is,

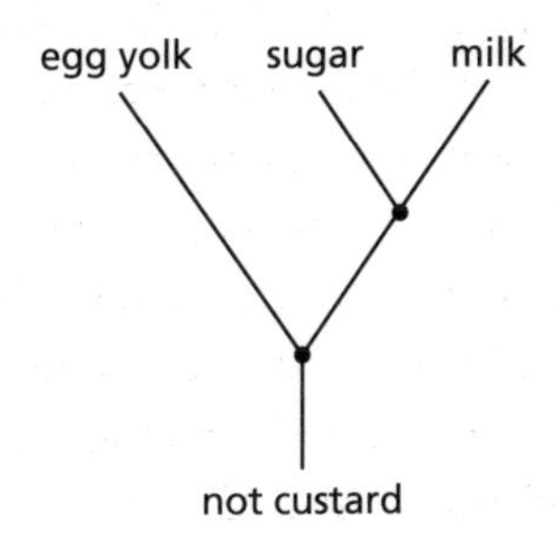

For the cake example, we have this version with four branches:

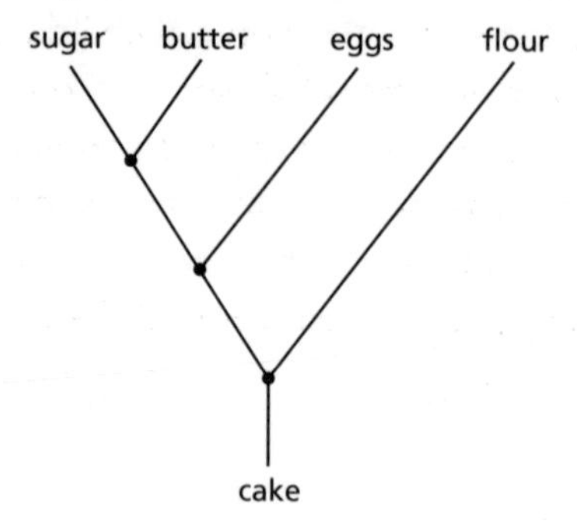

These diagrams are called 'trees' in maths, because they look a bit like trees. The loose ends at the top, which are labelled 'egg yolk', 'sugar' and 'milk' here, are called the *leaves* and the loose end at the bottom is called the *root*. They're another vivid way of bringing out the structure in a situation. Category theory studies these kinds of relationships carefully because it's something we take for granted in basic mathematical worlds that isn't true in other ones. This is the notion of associativity again. In the normal world of numbers, addition obeys this rule:

$$(5+5)+3=5+(5+3)$$

More generally, we can use symbols to show that this works for *all* numbers:

$$(x+y)+z=x+(y+z).$$

Now, what I've just shown with the custard is:

$$(\text{egg yolks}+\text{sugar})+\text{milk} \neq \text{egg yolks}+(\text{sugar}+\text{milk}).$$

Here the plus sign doesn't exactly just mean *plus*, and this is the whole point – it is a more subtle process of combining things than just throwing them together. And that's why the two versions aren't equal. If the process of combining those ingredients were much more crude, like 'chuck into a bowl together', then the two versions would be equal to each other, but they wouldn't very well resemble custard.

Category theory is well equipped to study situations a bit

better than the custard one, where the two versions of the tree are not *exactly* the same but 'more or less' the same, using the relationships that we are considering. This produces some interesting geometrical shapes as we'll now see.

We could try writing down all the possible trees with four leaves, like having four ingredients. Suppose we're only allowed to add one thing at a time. Here are all the possible trees:

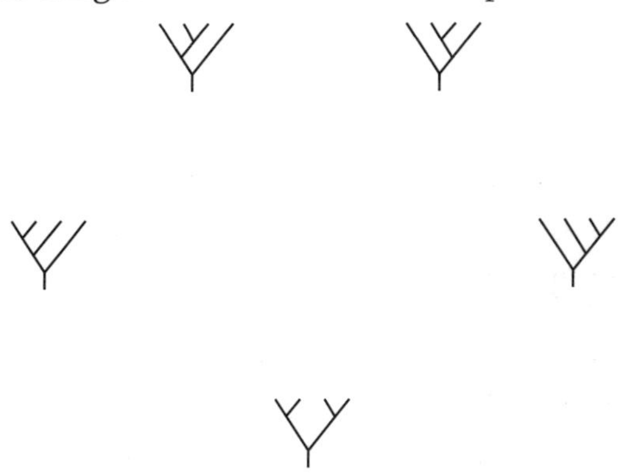

Now, to help us see the structure in this situation we can draw an arrow every time we have a branch moving its attaching point from left to right, because it's really a *process* of moving some brackets around:

Then we get this pentagon:

This is a very famous pentagon in category theory, which plays an important role whenever we're thinking about processes of putting things together in different combinations – which is very widespread in maths, whether it's by addition or by more and more subtle or complicated processes. Isolating the structure and drawing it like this as trees with arrows in between them means that we can turn a piece of *algebra* into a geometrical shape that neatly sums up all the information.

Moreover, we can play this game again (with some effort), and write down all the possible trees with *five* leaves. Then we can draw in the arrows again where we have a branch moving from left to right, and if we do it carefully we'll find that we have a three-dimensional shape with six pentagons and three squares. (This sounds tedious and long-winded but I admit it's the kind of game I love playing. I sat down and did it for all the trees with six leaves once as well.) You can cut it out and make it into a three-dimensional figure from this pattern:

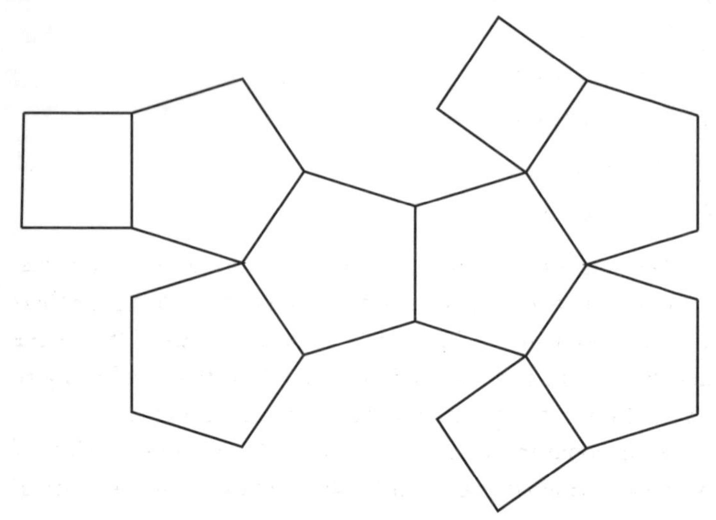

Just don't try to make it out of thick card, because it won't quite fit together – it needs to be made from paper that's a bit bendy, otherwise the pentagons and squares would have to be a bit wonky for them all to fit together.

And this all came from thinking carefully about how to understand the different possible processes of combining five ingredients. These shapes can be generalised to take account of more and more leaves, and of course the shapes become more and more complicated. Several fields of research deal with the problem of organising these complicated shapes.

Some things that may or may not be the same

Let's think about the set of numbers

$$\{1, 2, 3\}.$$

Which of the following sets do you think is sort of similar?

1. $\{2, 3, 4\}$
2. $\{2, 4, 6\}$
3. $\{-1, -2, -3\}$
4. $\{11, 12, 13\}$
5. $\{101, 102, 103\}$
6. $\{100, 200, 300\}$
7. $\{13, 28, 42\}$
8. {cat, dog, banana}

The first one is similar because all the numbers are just shifted up by one. The second is similar because all the numbers are multiplied by 2. The third is similar because it's just the negatives of the first set, the fourth and fifth are shifted up by 10 and 100 respectively, and the sixth is multiplied by 100.

What about the seventh? This is a rather random-looking set with no rhyme or reason to it. The eighth isn't even a set of numbers.

The important thing to notice here is that we naturally think about the *relationship* between the things in the set when wondering whether the sets are similar are not. But in fact, in mathematics a 'set' is just a bunch of objects where we have 'forgotten' about any relationships between them. So mathematically all these sets are 'the same' just because they have three objects each. This is not a very subtle notion of sameness, which is why in category theory we incorporate information about the relationships between things as well.

These sets are a situation where the wrong notion of 'sameness' made *too many* things the same. In other situations the wrong notion doesn't make enough things the same. For example, in the trees that we were looking at earlier on in the chapter, what mattered was how many leaves there were, and how the branches were connected up, not the angles at which they were connected, nor how thick the lines in question were. Sometimes notions of sameness are not as obvious as any of these things. What about this set of numbers?

$$\{13, 28, 41\}.$$

This looks quite a lot like the seventh set in the above list, but has a crucial difference – the third number is actually the sum of the first two, just like in the original set

$$\{1, 2, 3\}.$$

In the next chapter we'll see how situations like this can be expressed, where it's a relationship between several objects, not just between two, that is interesting.

14 UNIVERSAL PROPERTIES

Fruit crumble

Ingredients

50 g cold butter
50 g sugar (dark muscovado is good)

75 g flour

350 g fruit of choice, chopped if needed

Method

1. Mix the flour and sugar.
2. Chop the butter into small cubes and then rub into the dry ingredients with your fingertips, until it resembles bread-crumbs.
3. Put the fruit in an ovenproof dish with a little sugar if it seems necessary.
4. Cover thickly with the crumble mixture.
5. Bake at 180°C for 25–30 minutes until it looks brown and delicious.

Crumble is one of my favourite puddings. I love it because it's easy and comforting. I like the way that the crumble sort of blends in with the fruit on the surface, making a gooey layer between the crunchy crumble on top and the soft fruit underneath. My favourite fruit to use is blueberries. Or plums. Or bananas. You can basically use *any* fruit you feel like, although watermelon would be a bit strange. What about tomatoes?

At this point are you thinking 'But tomatoes are a vegetable' or 'Oh very funny'?

If you think tomatoes are a vegetable, you are characterising them by the *role* they generally play in our meals, rather than by their inherent characteristics. However, they are *technically* a fruit. What does that mean? It means that, according to the role they play 'in nature' as part of a plant's reproductive mechanisms, they are a fruit. However, if we used them as the 'fruit of choice' in the fruit crumble recipe, I suspect it would be rather bizarre. Tomato crumble is probably feasible, but surely only without all that sugar.

This is an example where in everyday language we characterise something by the role it plays in a particular context, rather than by its inherent characteristics. If you insist on referring to tomatoes as fruit all the time, or refuse to refer to peanuts as nuts because they're really a type of bean, then you are ignoring the context of these foods and the relationships they play with other food and with us.

Studying the role that things play is something category theory is well placed to do, because of the emphasis that we make on context and relationships. We have already seen that some things can be *completely* characterised by their relationships with other things. For example, the number 0 is the only number you can add to anything else without anything happening. This is a particularly special kind of relationship that category theory looks for, called a *universal property*.

Cinderella

The only person who fits in the shoe

When Prince Charming is looking for Cinderella, he doesn't go round asking people, 'Um, excuse me, are you Cinderella?' That would have made for a much less interesting story.

Instead, as we all know, he carries her glass slipper around (setting aside ongoing arguments about whether it's really supposed to be glass or fur) and gets everyone to try it on.

The key is that it's *tiny*, and so he knows that there's only one person whose foot could possibly fit into it.

He is looking for Cinderella according to some *characteristic* she has, rather than by her actual name – because he doesn't know her name. This is like referring to the prime minister as 'Prime Minister' rather than 'David Cameron' – you're referring to him by his role rather than by who he is as a person.

Category theory does this in maths. Because it's focusing on relationships with things, it seeks to characterise objects by their roles in relation to everything else. This is like playing the 'think of a number' game. Try this one:

I am thinking of a number.
If I add 1 to my number, I get 1.
If I add 2 to my number, I get 2.
In fact, if I add any number x to my number, I get x.
What is my number?

Or what about this one:

I am thinking of a number.
If multiply my number by 1, I get 1.
If I multiply my number by 2, I get 2.
In fact, if I multiply my number by any number x, I get x.
What is my number?

You have probably worked out that my first number was 0 and my second number was 1. These are very special numbers and they are characterised by what I just said in the 'think of a number' game. There isn't really another way of explaining what the number 1 is. Category theory makes this watertight.

But what about this one.

I am thinking of a number.
If I square it, I get 4.
What is my number?

Now, you probably worked out that my number could be 2.

But did you remember that my number could also be -2? The trouble with this one was that there was *more than one* possible correct answer. When Prince Charming went looking for Cinderella he was relying on the fact that there was only one possible person whose foot fitted the shoe. And in the 'think of a number' game, we rely on the fact that there's only one possible number that fits our description, otherwise we're not playing fair. Category theory seeks to characterise things in such a way that there can only be one possible answer, so that we've pinned down the role that this thing plays precisely.

If you think back to our axioms for numbers, we never actually said that there had to be *only one* possible number 0. This is because it is redundant as a rule – we can actually deduce it from the other rules in the following way.

We know that, for any number x,

$$0 + x = x.$$

Now suppose there's *another* number that behaves the same way as 0. Because it's trying to be another version of zero, let's call it Z. Now, because it behaves in the same way as 0, we know that, for any number x,

$$Z + x = x.$$

But because this is true for *any* number x, we can put in $x = 0$ and this gives us

$$Z + 0 = 0.$$

But we know that adding 0 to anything does nothing, so the left-hand side is Z, which gives us

$$Z = 0.$$

What we have shown is that this property of 0 characterises it uniquely, just like Cinderella's slipper – there is only one number that satsifies this property. It doesn't really matter what name we use for it (zero or nought, for example), as long as we know it satisfies this property, we must all be referring to the same number.

The same is true of *inverses*. Remember that the additive inverse of 3 is −3 because when we add them together we get 0. But in fact −3 is the *only* possible number with this property, which we can prove as follows.

Suppose there's some other number Y which also does this, so

$$3 + Y = 0.$$

But then we can add −3 to both sides (which amounts to subtracting 3 from both sides). On the left this gives us Y and on the right we get −3, so we have

$$Y = -3.$$

That is, if another number Y *tried* to be an additive inverse for 3, we would just discover it was −3 all along.

Finding a property that characterises an object *uniquely* is one of the important aspects of a universal property. Here 'universal' doesn't mean that the property holds universally for all objects. It's more like a universal key that works in all locks, or a universal password that you have on your computer to release all other passwords. It is in some way superlative with respect to all other objects.

Universal properties are like bests and worsts. Or firsts and lasts.

North Pole, South Pole

Looking at the extremities

The North Pole and the South Pole are fascinating concepts. The idea of actually going to the North or South Pole is one of the challenges to explorers who seek to conquer superlatives – climbing the highest mountain, for example. One fascinating thing about the North and South Poles is that there are no West and East Poles. This is because the earth is spinning in the east–west direction, not in the north–south direction; if it were spinning in the north–south direction we would have an East and West Poles instead, and all the magnetic fields would be another way up.

Studying the natural features of the poles helps us understand things about the world even though most of the world doesn't resemble the poles at all (thank goodness). There's a reason the only human settlements in Antarctica are scientific research stations.

Category theory also tries to find the 'North and South Poles' of each mathematical world, even if the rest of the mathematical world doesn't behave in the same way – these extremities give us insights into the rest of that world.

Once we know what the relationships between things are, we can look for different types of extremities like: which is the biggest/smallest or the strongest/weakest? For example:

* The smallest possible *set* of things is the empty set, which has nothing in it. It helps in maths to treat this a bit more actively than in normal life: it's like saying you have a stamp collection but it happens to be empty, rather than saying you don't have a stamp collection at all. Perhaps it's more like having an empty shopping trolley while you're at the supermarket, which is different from saying you don't have a shopping trolley.
* What about the biggest possible set? *Infinite* sets are very interesting, and it is amazing to try to compare different infinite sets and discover that some are 'more infinite' than others, in a very precise mathematical sense.

These are examples of 'universal properties'. They tell us something is special with respect to some relevant universe. We're not just saying something is big, which would be *a* property. We're saying it's the biggest, or some other mathematical version of a superlative. We're fascinated by finding superlative natural features of the earth – the tallest mountain, the deepest ocean, the longest river, the highest waterfall, and so on. It's a way of characterising our planet by extremity and of giving everything else on the planet a context. Category theory looks for the extremes of worlds even if they are not exactly typical – that's the whole point about being extreme.

> If we're talking about *groups*, the situation is a bit curious. You can't have a group with nothing in it, because one of the *axioms* for a group says it has to contain an identity object (the one that does nothing if you combine it with any other object). This is like the fact that you can't really have empty ravioli, because the whole point of ravioli is that it has something in it. Anyway this means that the smallest possible group is the one that has only one object in it, the identity object. When you combine it with itself you keep getting the same thing back again. This is like a number system containing only the number zero. It sounds silly, but we'll see later that it's quite important for *abstract* reasons even if not for *practical* reasons.

A less obvious but more mathematically important type of extremity is the 'initial object' and 'terminal object' of a category. Once we've drawn an arrow for each relationship in the category, we say an initial object is one that has exactly one arrow going out of it to *every other object* in the category. A terminal object has exactly one arrow going *into* it from every other object in the category. So initial objects are sort of at the 'beginning' if we think of arrows as being directional, and terminal objects are at the end.

This doesn't actually mean 'biggest' and 'smallest' any more than the North and South Poles are the biggest or smallest. It also doesn't mean best or worst. Remember this lattice of factors of 30:

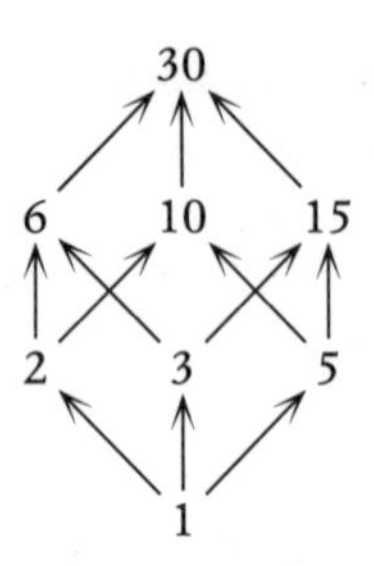

We can see from this picture that the biggest number is 30 and it is also *terminal*, and the smallest number is 1 and it is *initial*, but this is an accident of this particular example.

Actually you might find it easier to see that 30 is terminal from the picture with all the composite arrows drawn in:

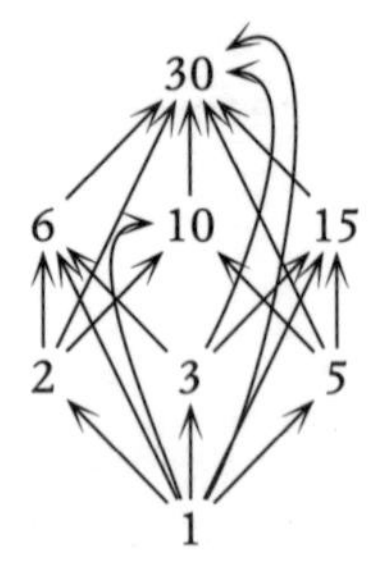

If you pick any other number in the picture, you can see that there's exactly one arrow going from it to 30. Likewise, there will be exactly one arrow going from 1 to your number, showing that 1 is initial.

For the category containing all possible sets and all possible functions between them, it turns out that the initial set is the empty set (which is the smallest), but the terminal set is any set with one object – definitely not the largest possible set.

The reason this is true is a bit technical. First of all we have to understand what 'arrows' we're thinking about here. The arrows in question are functions, where a function $A \longrightarrow B$ is a way of sending every object in the set A to an object in the set B. It doesn't have to be a process that can be written out as a neat-looking function like x^2

or something; it's more like a mysterious machine that takes objects of A as inputs, and spits out objects of B as outputs. If you open up the machine you might be able to see that it works according to a simple formula, but maybe it doesn't. Either way it doesn't matter *how* the machine does it; it just matters what the machine gives as the outputs.

Now, if B only has one object, there is only one possible machine – there's only one possible output, so no matter what object of A you feed it as the input, the output will always be the same, regardless of what convoluted process the machine goes through. So there is exactly one arrow from any set to B, making B terminal.

If A is the empty set, there is again exactly one possible machine. This time it's because there are no possible inputs, so the machine doesn't get to do anything at all. We've finished before we've even started.

As I said, universal properties often give rise to rather collapsed situations.

Big fish in a small pond

Moving to a different world to become an extremity somewhere else

If you want to be the biggest fish in a pond, you might have to move to a smaller pond, in which the bigger fish won't fit. If you're going to characterise something by a property rather than by its name, you had better make sure there's only one of them first. It would be hopeless to arrange to meet up with someone at 'the cafe in the National Gallery, London' because there are far too many of them, whereas meeting at 'the cafe in the Millennium Gallery, Sheffield' will work just fine. There was a bar in Chicago that my friends and I used to refer to as 'The Flamingo' even though it was actually called Bar Louie; the thing is that Bar Louie was a chain with branches all over the place, but there was only one building called The Flamingo, and only one bar in it.

My favourite whisky is Ardbeg Uigeadail, which for a while I referred to as the 'Ardbeg Unpronounceable'. I genuinely didn't know how to pronounce it, but discovered that if I asked for 'that unpronounceable Ardbeg' in a whisky shop, they knew exactly what I was talking about. However, there are now other unpronounceable Ardbegs such as 'Airigh Nam Beist' and so characterising the Uigeadail by this property no longer pins it down uniquely.

In category theory as well, if there is more than one thing with the same property, you can either move to a smaller pond or be more specific about the property you're talking about. You can tell this is going on in normal life when some superlative has too many qualifiers. For example, 'the best restaurant where you can have a three-course meal for under £15 in Sheffield' is not the best restaurant anywhere with a three-course meal for under £15, nor is it the best restaurant in Sheffield in all price ranges. Rather we're restricting *both* the property and the world in which we're considering it.

Sheffield has the distinction of being the 'largest city in England with no professional orchestra'; this has to be restricted to England as Glasgow doesn't have one either. Sheffield does have a symphony orchestra; it proudly calls itself the 'best amateur symphony orchestra in South Yorkshire'. My friends and I joke that I am the 'best young female category theorist in South Yorkshire'. Actually we could drop the 'best' and the 'young' – I am, to date, the *only* female category theorist in South Yorkshire, unless there are some secret ones hiding in Doncaster or something.

In the previous chapter we discussed the fact that sometimes we ask which objects are equivalent in a given context, but sometimes we start with an idea of which objects we want to think of as equivalent, and then *find* the context in which they are so. The same thing happens with universal properties. Sometimes we look for the objects that are universal in a given context, but sometimes we start with an idea of an object that is special, that seems like it *ought* to be universal somewhere, so we go round looking for the context in which it is universal, just like finding a smaller pond in which the fish is biggest.

We will later see how this works for some number systems: the natural numbers and the integers, for example. The natural numbers feel so 'natural' that it seems they should really be universal somehow. The same goes for the integers. In fact, mathematicians confusingly use the word 'natural' in this more hazy instinctive sense as well as in some very precise formal senses. If an object seems to a category theorist to be very naturally arising, then it seems somehow organic and unforced, and it seems like it should have a universal property *somehow*. The twelve-hour clock arithmetic is little contrived because we had to pick how many hours on the clock there would be, so it isn't really 'universal'. However, the number system with only a 0 in it seems a bit silly but it's still organic, because we didn't have to make any arbitary choices to make it happen. Likewise, the integers just spring up without us really doing anything. But the integers are neither initial nor terminal among number systems, because it turns out that the 0 number system is both initial and terminal. So we have to find another context in which the integers are universal, and we'll see what it is soon.

Here's why the smallest possible group is both initial and terminal. This is also a somewhat technical fact, and it involves understanding what is the relevant notion of relationship between groups. The answer is that an arrow $A \longrightarrow B$ is a way of sending every object of A to an object of B (just like with functions) with the added axiom that the notion of addition has to work sensibly once you've moved to B. What that means is that if you send an object a_1 to b_1 and an object a_2 to b_2, then $a_1 + a_2$ has to go to $b_1 + b_2$.

One consequence of this is that the identity in the group A has to be sent to the identity in the group B. And so, if the group A is the one that *only* has the identity in it, you have no choice about where to send it – it simply has to go to the identity in the group B. So there is precisely one arrow from A to B no matter what B is, which means that A is initial.

However, there is also precisely one arrow from B to A, if A is this group that only has the identity in it. Because you have no choice where to send everything in B, just like with the example of the set with only one object. This means that A is both terminal and initial. It's a bit like being in a world where the North and South Poles are the same.

Big garden

When being superlative is a burden

Sometimes being the biggest isn't always the best. Having a big garden might sound nice, but it would require an awful lot of gardening; of course, this would be fine if you were rich enough to pay for a team of gardeners. Having a big car might sound nice, but it's also much more unwieldy to manoeuvre, unless you're in the US where everyone else also has a big car, so the roads are wider and the parking spaces are bigger. Being extremely tall might help if you're a basketball player or trying to change a light bulb, but it's not so great when you're trying to stuff yourself into an aeroplane seat.

For many things in mathematics there's a trade-off between being the 'biggest' and being the 'most practical'. The 'biggest' ones are good in theory – it's illuminating to think about them, and to put other things in context. But once that context has been found, the aim is often to find more 'usable' versions for daily mathematical life.

For example, the twelve-hour clock is not universal, because we've imposed a contrived rule on it that every time we get round to 12 we act as if we're at 0 again. However, for practical purposes this is much better. Imagine if we never imposed a rule saying we went back to 0 again? We would have to say things like 'I'll meet you at half past twenty-nine million six hundred and twenty seven thousand four hundred and seventy-three'. This would be if we told the time using all the natural numbers, rather than the twelve-hour clock version. The natural numbers are *universal*, but the twelve-hour clock is *practical*. Things that are universal are good for abstract thoughts. After all, we never actually need *all* the natural numbers in daily life, we just need to know that we will never run out in principle.

But we still need to know what this universal property is that encapsulates the fact that the natural numbers arise organically by just counting and counting and counting forever. We're nearly ready to see what this is.

Erdős

• • • • •

When minimalism helps us see what's what

Apparently everything in Paul Erdős's life was in service of his mathematics, and he owned nothing and did nothing that was extraneous to this purpose. He hardly had any possessions, and he rarely stayed very long in any one place, travelling around with his suitcase to discuss mathematics with different people in different places. He would turn up somewhere with his suitcase, discuss maths with someone for some days or weeks, and then move on to the next place where he wanted to discuss maths.

Category theory often seeks to characterise things by what role they play, but it also does it the other way round as well: it thinks up a role, and then goes round looking for something that plays that role in the most minimal possible way, without any extraneous features. Because then not only does the role characterise the thing, but the thing characterises the role as well. It's like the fact that Harry Potter has only ever been played by Daniel Radcliffe, and for a while Daniel Radcliffe had only ever played Harry Potter. Until he appeared in *Equus*, Harry Potter *was* Daniel Radcliffe and Daniel Radcliffe *was*

Harry Potter. By contrast, James Bond has been played by many actors, but people love to argue about which one is the 'definitive' James Bond.

There are many composers who only wrote one violin concerto: Tchaikovsky, Mendelssohn, Brahms, Beethoven, Sibelius, Bruch. So we can say 'the violin concerto by Tchaikovsky' (or any of the others) without ambiguity, whereas 'the violin concerto by Mozart' could refer to many different pieces, and 'the violin concerto by Schubert' would be referring to something non-existent.

But most of these composers also wrote other famous works – apart from Bruch. Bruch basically only wrote a violin concerto. This isn't actually true, but the violin concerto is the only thing he wrote that's really famous. So not only is his violin concerto defined by being written by him, he is also somewhat defined by his violin concerto.

The category theorist James Dolan likens all this to a guy walking up the street whose moustache is so enormous he is completely dominated by it – in fact, the person seems to exist only as a carrier of the moustache. He's a 'walking moustache'.

Category theorists often refer to such minimal features as 'free-living'. Imagine breaking free of all constraints and just living with the bare minimum of what was necessary to sustain life. (A friend of mine ran away from home at the age of 16 and took her parents' blender with her: a key necessity to sustain life?)

Understanding the bare minimum of what something needs to sustain its life is a key feature of category theory. In this sense, Erdős truly was a 'free-living' mathematician, living only with those things necessary to sustain his mathematical life. He was in fact a 'walking mathematician' in both this figurative sense, and the literal sense, walking from place to place with his minimal suitcase.

Putting the 'natural' in the natural numbers

This leads us to seeing what is universal about the natural numbers. The answer bears a pleasing resemblance to our intuition that they are what you get naturally if you start with 1 and just keep counting forever.

In category theoretic terms, this is called 'free'. It means that you start with something and proceed freely, and never impose any extra rules on yourself apart from the ones that automatically come with the context you're in.

The context for natural numbers is the notion of a 'monoid'. This is something that is like a group, so we can add things up in any order we want, but without the rule saying that everything has an *inverse*, so we don't worry about negative numbers. Now, if we start with just the number 1 and make a monoid 'freely', we know that we have to be able to do

$$\begin{gathered} 1+1 \\ 1+1+1 \\ 1+1+1+1 \\ \vdots \end{gathered}$$

We know that it doesn't matter how we put brackets around these things, but we are not going to impose any more rules on ourselves, because we want to be free. No rules. This means that we will never get any extra equations saying things like

$$1+1=1+1+1+1$$

or anything like that. So all we do is keep adding ones, and what we get are the natural numbers. So the natural numbers are the *free monoid* starting with just the number 1.

If we demand inverses as well, so that we have a *group*, then starting from just the number 1 gives us all the integers. Because basically all we can do is add up 1's as above, and then take the negative versions as well. So the integers are the *free group* starting with just the number 1.

In category theory we can make free objects starting from other things as well. We can make a free group starting from any set of things. The freedom of this situation is a type of universal property that is closely related to 'forgetting structure' as we discussed in the chapter on structure. We saw that there was the idea of 'forgetting' the structure of a group to get a set, and now we have the notion of 'freely' building up a group starting from a set. Likewise, we thought about *rings* which are like groups but have multiplication as well as addition. We saw the notion of 'forgetting' the multiplication involved in a ring to get back a mere *group*, and in fact there is also the notion of 'freely' building a ring starting from a group. The processes of forgetting things and building them freely are a type of opposite, but they're not actually inverse to one another. They're another special type of relationship that category theory looks at that's even more subtle.

Exploring more universal properties

$1 + 1 = 2$, or does it?

Sometimes when I tell people I'm a mathematician they make jokes about one add one being two. Either they tell me that's all the maths they're really sure about, or they tell me that maths is all either right or wrong, because, for example 'one add one just does equal two, end of story'.

Of course, we've already seen a place where $1 + 1 = 0$, on the two-hour clock. Let's see a way that this clock idea arises really.

Let's start by changing the question a bit: $7 + 7$ just *does* equal 14, doesn't it? Well yes, unless you're working on a twelve-hour clock, in which case 7 o'clock plus 7 hours is 2 o'clock.

$$7 + 7 = 2.$$

But we're trying to think about something other than clocks. What if you're thinking about days of the week. This works better in Chinese, where Monday is called 'day one', Tuesday 'day two', Wednesday 'day three', and so on. (Don't be fooled though: Sunday is called 'day sun'.) Anyway, if we're on 'day five' and we add three days, we get to 'day one'.

$$5 + 3 = 1.$$

Or what if we're playing a piece of music and we're thinking about beats in a bar. Say there are four beats in the bar. Then two beats after third beat in the bar is the first beat of the next bar:

$$2 + 3 = 1.$$

Now you might be tempted to argue, 'This doesn't count!', which is not a bad mathematical response. Mathematicians often just declare that things don't count if they don't fit into their world. However, mathematicians only say 'this doesn't count' temporarily. If something doesn't fit with a world, but still makes some kind of sense, they say it doesn't count in *this* world, but then they go and look for the world in which it *does* make sense.

All of these 'weird' addition laws do make some kind of sense. They're a lot like our normal number system – in fact, they're enough like our normal number system that they count as just another kind of number system. That is to say, we could check that they satisfy the axioms for numbers that we came up with before – the order of addition doesn't matter, brackets don't matter, there's a number that acts like 0 and there are numbers that act like negatives.

What about counting 'not's? Children discover that 'not's cancel each other out and get very excited about making silly jokes like 'I'm not not hungry' meaning that they are hungry. Or saying 'I'm not not not not not not not not not not not not hungry!' and then collapsing into giggles because they know nobody has a chance of working out whether they said an even number of 'not's, meaning that they are hungry, or an odd number, meaning that they are not hungry.

In this case,

$$\text{not not hungry} = \text{hungry}$$

or we could say:

$$1 \text{ not} + 1 \text{ not} = 0 \text{ nots.}$$

There: $1 + 1 = 0$.

This is a perfectly valid number system, and moreover, it arises naturally and is pretty useful. In this number system, there are only two numbers, 0 and 1. And you add them up like this:

$$0 + 0 = 0$$
$$0 + 1 = 1$$
$$1 + 0 = 1$$
$$1 + 1 = 0$$

As we saw in the previous chapter, we could draw this in a little addition table, like this:

+	0	1
0	0	1
1	1	0

and it's the same pattern as in a Battenberg cake:

You get the same pattern if you think about NOT gates in electric circuits, or light bulbs that have a light switch at two different locations in the room – if you switch only one switch, the light will go on, but if you switch both switches, it will go off again.

This is a pretty small number system. But is it the smallest possible? No, in fact there's an even smaller one with only one number: 0. This has an addition table like this:

+	0
0	0

This is like a world in which you're not allowed any sweets, ever. Like me when I was little as I was so allergic to food colouring (and all sweets had food colouring in those days). The only number of sweets that existed in my world was 0, and we have got ourselves back to the *smallest possible group* in which there is only one object: the identity. Remember, the identity (if we're thinking about addition) is 0, because when we add it to anything else, nothing happens.

This is not a very useful number system to think about all by itself, but in category theory we don't just think about number systems by themselves – we think about relationships between number systems.

When I was little I compared my sweetless world with all my friends who got 5p or even 10p each Friday and could go down to the village sweet shop and buy quite a giant bag of sweets for that princely sum. Similarly in category theory we compare this numberless number system with all the other number systems, and it is the South Pole of the world of number systems. It is the extreme number system, in which nothing much can happen (like at the South Pole) but which is still an important thing to pin down, as it tells us where the extremity of our world is.

Extreme notions of distance

We've also looked at notions of distance, called 'metrics'. There's a most extreme possible version of a metric where *everything* is a distance 1 from everything else (unless they're equal). For these abstract distances we don't use units, so it's not 1 km or 1 mile, it's just 1 *something*. In this way a metric where everything was a distance 10 apart wouldn't be any 'bigger', because we don't have any units, so '1 somethings' is abstractly the same as '10 somethings'. The point here is that everything is unavoidably separated from everything else. This notion of distance sounds a bit silly, but we can check that it satisfies the three rules for a metric.

1. The distance between A and B is only 0 if A and B are the same (because, after all, the distance is otherwise 1).
2. The distance from A to B is the same as the distance from B to A (because either they're the same, in which case the distance is 0, or they're different, in which case the distance is 1).
3. The triangle inequality – this is a bit more complicated to check, but it does still work.

If we write the distance from A to B as $d(A,B)$, then we need to show

$$d(A,C) \leqslant d(A,B) + d(B,C).$$

We can draw a table with cases:

	$d(A,B)$	$d(B,C)$	$d(A,B) + d(B,C)$	$d(A,C)$
$A = B = C$	0	0	0	0
$A = B \neq C$	0	1	1	1
$A \neq B = C$	1	0	1	1
$A \neq B \neq C, A \neq C$	1	1	2	1
$A \neq B \neq C, A = C$	1	1	2	0

What we have to check is that the last column is always less than or equal to the second-to-last column, which indeed it is.

Another way of proving that this inequality is true is by doing a *proof by contradiction*. Suppose there's some A, B, C where the inequality is false, so

$$d(A,C) > d(A,B) + d(B,C).$$

Our aim now is to 'hope for the worst', or rather, discover that this implies some sort of contradiction, so can't be true.

Now, all the distances are 0 or 1, so the left-hand side can only be 0 or 1, and the right-hand side can only be 0, 1 or 2. So the only way the left-hand side can be bigger than the right-hand side is if the left-hand side is 1 and the right-hand side is 0. But the only way the right-hand side can be 0 is if both distances on the right are 0, which means $A = B = C$, which means the left-hand side is 0. This makes the two sides equal, which contradicts our assumption.

Which of these two arguments did you find easier to follow? Which was more satisfying?

This metric is called the 'discrete metric', because it makes things all spaced out in discrete bits. Nothing is very close together, but everything is equally far apart. (Perhaps this is a teleportation metric, where every place is equally easy to get to?) The fact that this metric seems a bit absurd is not unusual for things with universal properties – they are extreme examples of things, and so are either very collapsed on themselves or very stretched out.

You might wonder if there's a 'smallest possible metric' in which the distance between everything is 0. The answer is yes, except that this would mean everything would have to equal everything else as well. Again, very collapsed in on itself, instead of stretched out.

Extreme notions of category

You might be wondering if there are extremities in the land of categories itself. The answer is yes.

The smallest possible category is the empty one, just like the smallest possible set. And just like the smallest possible set, this is *initial* in the *category of categories*. The terminal category is the one with exactly one object, and exactly one arrow, that we saw before:

It's like a conflation of the terminal set and the terminal group.

This is because the relevant notion of relationship *between categories* is a conflation of the notions we've seen for sets and for groups. To get from one category to another we must not only send every object to an object, but we must also send every arrow to an arrow, and composition has to work sensibly just like addition had to work sensibly when we were doing this for groups. So if we're trying to get from any category A to the little one above, then we have no choice at all – every object of A has to go to the single object of x, and every arrow of A has to go to the identity arrow. So this little category is *terminal.*

To understand this more, imagine that we're trying to send the triangular category to the little one:

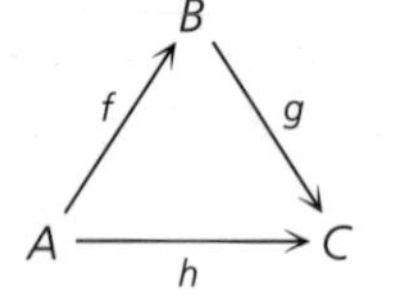

We have to feed in A, B, C and f, g, h as inputs, and the outputs have to be x or the identity morphism. When we feed in an object as input, we have to get an object as output. This means that when we feed in A, B or C, we *have to* produce x as the output. And when we feed in a morphism as input, we have to get a morphism as output too, so when we feed in f, g or h, we *have to* produce the identity morphism as output. Hence there is only one possible 'function machine' from the bigger category to the little one. The same sort of argument will still work no matter how big the first category is. This shows that the little one is terminal.

There are also universal versions of categories that are a bit like the discrete metric we discussed above, in which the distance from everything to everything else was 1. The category version of this is a 'discrete category' in which there is no arrow from anything to anything else – all the objects are completely separate from each other.

However, unlike for metrics, we can have a sort of 'opposite' notion of this category, in which everything is related to everything else, without actually being equal. In this category there is exactly one arrow from every object to every other object. This is called the 'indiscrete' category. Note the spelling – it's not indiscreet (divulging all secrets) but *indiscrete*, a word that is not used much outside maths. It means that rather than the objects all being very separate, the opposite is true – the objects are not separate at all. This doesn't mean that they're all identical, but it does mean that they're all equivalent in this particular context. The chart of the very tightly knit group of friends in the chapter on relationships is an example of an indiscrete category. The friends in this picture aren't identical, but they're equivalent in the sense that perhaps they all know the same things about each other's lives; you know, the kind of group of friends where if you tell something to one of them, you have effectively told all of them.

Finding a universal property in category theory not only tells you something important about the object in question, but it means you can look for things that have the same sort of universal property in other contexts, and it gives you an interesting point of comparison between those worlds. It also gives you access to one of the things mathematics is so keen on

– the possibility of studying a diverse range of things at the same time by finding a way in which they're all similar.

Here are some mathematical examples of things that turn out to be comparable through their universal properties.

* Adding numbers up can be seen in the same light as taking unions of sets, that is, making a new set consisting of all the objects in the previous two. Likewise, so can highest common factors of numbers, or the surfaces you get when gluing two surfaces together. These are all a type of *colimit*, which means they have a particular kind of universal property.
* Multiplying numbers can be seen in the same light as Cartesian coordinates (an *X*-coordinate and a *Y*-coordinate), or taking the maximum or minimum of two numbers, or making a doughnut shape by swooping a circle through the air as we saw earlier on, or iterating a Battenberg cake.
* The natural numbers can be seen in the same light as the integers, but by contrast we *can't* see the real numbers in this light; they're genuinely different.

The natural numbers and integers are both *freely generated* structures. We can generate the natural numbers from the number 1, by adding it repeatedly. We can generate the integers from the number 1 by adding and subtracting it repeatedly. But there's no way of generating the real numbers from one number and some operations – you're doomed to miss out some of the real numbers, even if you start with an irrational number.

Here is category theory's way of looking at this. The natural numbers form a monoid, with addition. The integers form a group, with addition. The real numbers form something called a field – there's addition, subtraction, multiplication, and division for everything that isn't zero. The thing is that the category of all monoids and the category of all groups both have good universal objects inside them, whereas the category of all fields does not.

Universal properties give us a clue for how we should move from one world to another when making a mathematical correspondence. Just as the prime minister of the UK is more or less analogous to the president of the US, we look for corresponding universal objects in different mathematical worlds in order to understand the relationships not just between objects inside the worlds but between entire worlds themselves.

Some of the examples in the lists above seem much more obviously similar to each other than others. One of the satisfying things about category theory is that you can keep getting more abstract until more and more things become 'the same' and can be studied together. In fact there's a joke about this among category theorists, which comes from a comment in the great *Categories for the Working Mathematician* by one of the subject's founders, Saunders Mac Lane:

All concepts are Kan extensions.

A Kan extension is something with a certain universal property. Mac Lane's assertion is that not only can everything be understood via some universal property or other, but everything can be understood via *the same* universal property. This is a rather grand unifying vision of mathematics. Although it's something of a joke, it also sheds light on what category theory is at heart.

15 WHAT CATEGORY THEORY IS

We said, in the first half of the book, that mathematics is there to make difficult things easy. We have now seen that category theory is the mathematics of mathematics. So, category theory is there to make difficult *mathematics* easy.

In the second half we have discussed various ways in which it goes about this, but I want to conclude by characterising category theory as a category theorist would: what is the glass slipper that fits category theory exactly? That is, instead of saying what category theory looks like, we're going to say what role it fills.

Truth

People often think that mathematics is all either right or wrong. That's not true – even if a piece of maths is right, it can still be good or bad, it can be illuminating or not, it can be helpful or not, and so on.

However, there's a grain of truth in this business of right and wrong. One of the remarkable qualities of mathematics is that, because it's all built from logic and nothing else, mathematicians can readily agree when something is right. This is very different from other fields, in which opposing theories can be argued forever. As philosopher Michael Dummett remarked in *The Philosophy of Mathematics*:

> Mathematics makes a steady advance, while philosophy continues to flounder in unending bafflement at the problems it confronted at the outset.

Mathematical fact has an elevated status over other kinds of fact. We've already discussed the fact that scientists revere the so-called scientific method, experimental method and evidence-based knowledge, where facts are deduced from hard evidence that can be experimentally repeated. Maths isn't like that at all – it doesn't use *evidence*, because evidence isn't logically watertight. Evidence is the foundation of science, but it isn't enough to give us mathematical truth. This is why mathematics is sort of a part of science, but also isn't a part of science.

Mathematics uses the 'logical method', where facts are deduced only using cold, bright logic. Mathematical truth is revered because of proof: everything is rigorously proved, and once it has been proved, it cannot be refuted. You can find a mistake in a proof, but that means it was never really proved in the first place. Thanks to the notion of 'proof', we have an utterly unassailable way of knowing what is and isn't true in mathematics. How do we show that something is true? We prove it.

Or do we?

The wonderful thing about formal mathematical proof is that it eliminates the use of intuition in an argument. You don't have to guess what someone is trying to say, or interpret their words carefully, or listen to the inflection in their voice, or look at the expression on their face, or respond to their body language. You don't have to take into account the nature of your relationship with them, the stress they're under at the moment, the fact that they might be drunk, or the way their past experiences might be affecting them now. You don't have to be able to imagine what something looks like, you don't have to be able to imagine eight-dimensional space, or what a pile of two million apples would be like, or how it feels to be at the North Pole. All of these problematic subtleties are gone.

And the trouble with formal mathematical proof is that all of these subtleties are gone. The subtleties that can cause problems are also useful, but useful for something different. They are useful for getting a personal insight into something. You might think that mathematics shouldn't be about personal insight, but in the end *all* of understanding is about personal insight. It's the difference between understanding and knowledge. Formal mathematical proofs may be wonderfully watertight and unambiguous, but they are difficult to understand.

Imagine being led, step by step, through a dark forest, but having no idea of the overall route. If you were abandoned at the start of that route again, you would not be able to find your way. And yet, when you're led there step by step, you do make it to the other side.

Mathematicians and maths students have all had the experience of reading a proof and thinking, 'Well, I see how each step follows from the previous one, but I don't have a clue what's going on.' We can read a correct proof, and be completely convinced of each logical step of the proof, but still not have any understanding of the whole. Here's a completely formal proof of a very trivial-sounding fact: *any statement implies itself.* Note that by 'implies' here we mean logical implication. In mathematical logic 'implies' doesn't mean quite the same as in normal life – it means something much stricter. 'A implies B' means that if A is true then B is *definitely* true without any room for doubt. In normal life we say things like 'Are you implying that I'm stupid?' and implication is more of a suggestion or an insinuation, not a hard and fast fact.

Back to our example of statements implying themselves. This is a bit like things equalling themselves – the most obvious equation is

$$x = x.$$

Surely this is true about logical implication as well?

For example:

* If I'm a girl, then I'm a girl.

* If it's raining, then it's raining.
* If $1 + 1 = 2$, then $1 + 1 = 2$.

And yet, look how absurd and convoluted the rigorous proof of this is. Here the little arrow sign means 'implies', and this is the completely rigorous proof that any statement p implies itself, using the axioms of formal logic.

Proof of ($p \Rightarrow p$)

$$(p \Rightarrow ((p \Rightarrow p) \Rightarrow p)) \Rightarrow ((p \Rightarrow (p \Rightarrow p)) \Rightarrow (p \Rightarrow p))$$
$$p \Rightarrow ((p \Rightarrow p) \Rightarrow p)$$
$$(p \Rightarrow (p \Rightarrow p)) \Rightarrow (p \Rightarrow p)$$
$$p \Rightarrow (p \Rightarrow p)$$
$$p \Rightarrow p$$

I admit that I find this proof extremely exciting and satisfying, but not even all mathematicians will agree with me. I only included it here so that you could marvel at how ludicrously complicated it seems to be to prove the most basic logical statement. Non-mathematicians think they'll never understand what mathematicians do, but half the time mathematicians don't understand each other either. Does this proof convince mathematicians that any statement really does imply itself? No, of course not.

So, if the proof by itself doesn't convince them of the truth, then what does?

The trinity of truth

There is something else that plays the role of convincing mathematicians that something is true. I think of it as an *illumination*.

I'm going to talk about three aspects of truth:

1. belief
2. understanding
3. knowledge.

This is a bit like the three domes of St Paul's Cathedral. We have knowledge, which is what the outside world sees, belief, which is what we feel inside ourselves, and understanding, which holds them together.

The interplay between these three types of truth is complex. We can start by drawing a Venn diagram for them:

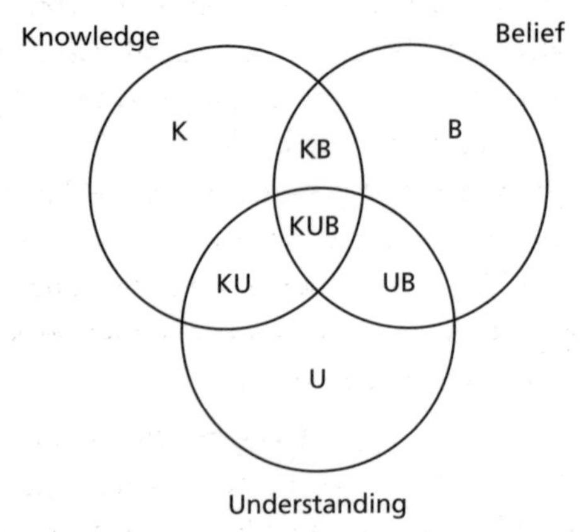

I've marked the different areas of overlap, so we have:

KUB: Things we know, believe and understand. The most secure of truths.

KB: Things we know and believe, but do not understand. This includes scientific facts that are certainly true, even if we don't understand them. For example, I don't really understand how gravity works, but I know and believe it works. I know and believe that the earth is round, but I don't understand *why*.

B: Things we believe, but do not understand or know. These are our axioms, where everything else begins –

the things we can't justify using anything else. For example, for me, there are things like love and the preciousness of life. I believe that love is the most important thing of all. I can't explain why, and I can't say I know for sure it is true – because what does that even mean?

After this things get a bit trickier.

K: Things we know, but do not understand or believe. Is this at all possible? I think if you've ever known sudden grief or heartbreak you might know what this is like. Those numb days after the event when you know, rationally, that it really has happened, but you simply can't believe it, you can't feel it to be true in your stomach. And you certainly don't understand it. Perhaps extremes of good emotions feel like too. Perhaps if I won the lottery I would, for a while, know that it had happened without understanding or believing it. Winning the lottery of love feels like that too, at the height of its ecstasy.

KU: Things we know and understand, but do not believe. Perhaps this is where we get to the next stage of grieving, when we have come to understand that this terrible thing really has happened, but we still don't believe it. But if you're in this state you're probably in some state of denial, because usually knowing and understanding something would make you really believe it's true.

Finally we have the following sections which I suspect are empty.

U: Things we understand, but do not know or believe.

UB: Things we understand and believe, but do not know.

I don't think it's possible (or rather, reasonable) to understand something without knowing it. In this way, understanding is different from the other two forms of truth, which do seem to be able to exist by themselves. Truth flows through this diagram in one direction only – from understanding flows everything else.

Of course, it all depends somewhat on exactly how we define these things, but just try thinking for a second about some things you believe. Here are some things you might believe:

- $1 + 1 = 2$.
- The earth is round.
- The sun will rise tomorrow morning.
- It is very cold at the North Pole.
- My name is Eugenia.

Why do you believe these things? Perhaps you think you understand why $1 + 1 = 2$, except when it doesn't, as we've discussed earlier. If we are working in the natural numbers or integers, $1 + 1 = 2$, mostly because that's the *definition* of the number 2. But $1 + 1 = 0$ if we're working in two-hour clock arithmetic, that is, the integers modulo 2.

But why is the earth round? Why will the sun rise tomorrow morning? Why is it cold at the North Pole? These are things that most of us know, but without really understanding them. I think a lot of our personal scientific knowledge is just that – knowledge that we believe because somebody we trust has told it to us. We have taken it on trust, or on authority.

Why is my name Eugenia – if it is? That last one is fairly easy, assuming that is my name: it is so because my parents chose it. But are you going to believe that, just because it's on the cover of this book? Or would you have to go and look up the record of my birth before believing it? (I hope not.) This is more complex. You might believe it's true without really knowing if it's true or not.

Understanding is a mediator between knowledge and belief.

In the end the aim is to get as many things as possible into the central part of the picture, where knowledge, understanding and belief all meet.

Here's a mathematical example of the difference between knowledge and understanding. Suppose you are trying to solve the equation

$$x + 3 = 5.$$

Perhaps you remember that you can 'take the 3 to the other side and switch the sign'. So the next step is

$$x = 5 - 3$$

and we see that x is 2.

However, knowing that this works is not the same as understanding it. Why does it work? It's because we have an equality between the left-hand side and the right-hand side, and so we can do the same thing to both sides and they'll still be equal. Now, we want to get the x isolated by itself on one side, which means we want to get rid of the 3 on the left. How do we do that? We subtract 3. But if we do that on the left we have to do that on the right as well. So what we're really doing is

$$\begin{aligned} x + 3 &= 5 \\ x + 3 - 3 &= 5 - 3 \\ x &= 2. \end{aligned}$$

Understanding this *principle* rather than merely knowing the rule makes the knowledge more transferable to other situations.

Pickpocket/putpocket

Remember the strange case of the 'putpocketing' from Chapter 4? You had a ten-pound note in your pocket. Someone pickpocketed you, but also someone else slipped a ten-pound note into your pocket afterwards. So you believe you have a ten-pound note in your pocket.

But do you actually know you do? Perhaps you then check to see if your ten-pound note is still there. At this point, you now also know you have a ten-pound note in your pocket.

But until someone enlightens you about the whole story, you will not actually understand *why* you have a ten-pound note in your pocket.

Why? Why? Why?

Why did the chicken cross the road?

The key to understanding is the question, Why? Why is such-and-such true? 'Because we've proved it' is not a satisfactory answer, from a *human* point of view. Why is that glass broken? 'Because I dropped it' or 'Because the molecular bonds between the glass molecules are no longer in place'. We've all heard, 'We apologise for the late departure of this flight. This is due to the late arrival of the incoming flight . . .' And, of course, why did the chicken cross the road? Asking why is like asking what the moral of the story is.

Let us try asking some mathematical whys.

1. Why is the area of a triangle half the base times the height?
2. Why is minus minus one equal to one?
3. Why is zero times anything zero?
4. Why can't you divide by zero?
5. Why is the ratio of the circumference of a circle to its diameter always the same (it's π)?
6. Why does the decimal expansion of π go on forever?

Let's try answering these now. The area of a triangle is quite easy to think about if it's a *right-angled* triangle, because then the triangle is obviously half of a rectangle:

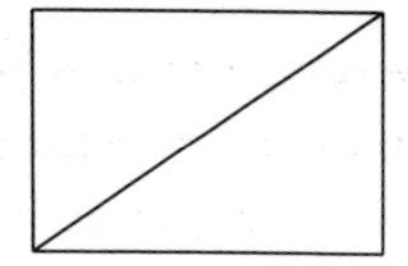

If it's a more random-looking triangle, like this one:

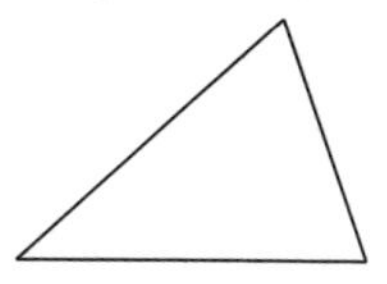

then we have to fill it in to a rectangle a bit more cleverly, say like this:

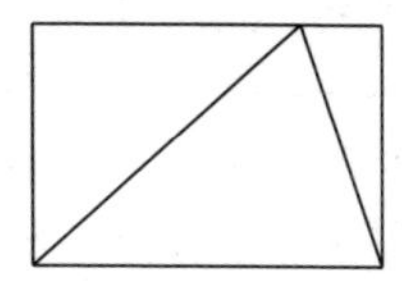

and then work out why the extra parts can be pieced together to make the same triangle that we started with:

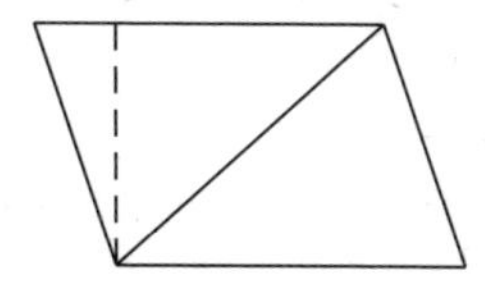

That's pretty convincing, but it's not quite a proof.

For the next one, we could do a proof using the axioms for numbers. Formally it looks like this.

The additive inverse of x is defined to be $-x$, that is,

$$-x + x = 0$$

and it is unique with this property. We need to show that 1 is the additive inverse of -1. That is,

$$1 + (-1) = 0.$$

But this is true since -1 is the additive inverse of 1.

This is mathematically correct, but not exactly *convincing*. Are you more convinced if I say something like 'putting a minus sign flips which way we're facing, and if we flip twice we get back to the direction we started'? Not mathematical at all, but possibly more convincing. Perhaps it would be more convincing to put it like this. Whenever we have $a + b = 0$, this tells us that a and b are additive inverses of each other, that is,

$$a = -b \quad \text{and} \quad b = -a.$$

We know that -1 is the additive inverse of 1, so we can put $a = -1$ and $b = 1$ and we get $a + b = 0$. Now we can conclude that $b = -a$, which in this case means

$$1 = -(-1).$$

This is essentially the same proof as before, but written out a bit less elegantly. Did you find it more convincing?

As for multiplying by 0 giving 0, there is a similarly technical and even more unilluminating proof from the axioms that looks like this:

Let x be any real number. Then

$$\begin{aligned} 0x + 0x &= (0+0)x && \text{distributive law} \\ &= 0x && \text{definition of 0} \end{aligned}$$

Subtracting $0x$ from both sides, we get $0x = 0$.

We have already discussed the fact that 'you can't divide by 0' really means '0 has no multiplicative inverse according to the axioms'. But with all of these proofs from the axioms for the real numbers, we are not really trying to justify *why* these things are true. Rather, the proofs are only to check that the things we *feel* are true really are true according to the axioms we've chosen. It's not actually an explanation of anything.

The fact about circles can be proved using calculus, but you can also try to convince yourself like this: both the circumference and the diameter are *lengths*, and when you scale a shape up or down all its *lengths* stay in proportion.

As for the decimal expansion of π going on forever, you might remember it's because π is irrational. But why is π irrational? I don't know of a particularly convincing *explanation* of that, except that circles are curved, and diameters are straight, and it would seem a bit oddly neat and tidy if the ratio was something rational.

Actually some rational numbers have decimal expansions that go on forever too, such as $\frac{1}{9}$, which is 0.1111111… However, the decimal expansion of a rational number always ends up repeating in cycles, whereas the decimal expansion of an irrational number like π or $\sqrt{2}$ never repeats itself.

You can always keep on asking why, because there is always another level of 'why' that can be asked. Every child knows that the question why is actually an infinite sequence of questions with which to harass an adult.

The point of the above examples was to illustrate the fact that if you ask *why* a mathematical fact is true, the mathematical proof is often not something that will convince you why it is true. Instead, it might convince you *that* it is true. And there's the crucial difference.

Proof vs illumination

Proof has a sociological role; illumination has a personal role.

Proof is what convinces society; illumination is what convinces us.

In a way, mathematics is like an emotion, which can't ever be described precisely in words – it's something that happens inside an individual. What we write down is merely a language for communicating those ideas to others, in the hope that they will be able to reconstruct the feeling within their own mind.

When I'm doing maths I often feel like I have to do it twice – once in my head, and then a second time to translate it into a form that can actually be communicated to anyone else. It's like having something you want to say to someone, which seems perfectly clear in your head, but then you find you can't quite put it into words. The translation is not a trivial process; why do we try to do it at all? Why do we not just stick to the things that are illuminating?

- (a) Illumination is very difficult to define.
- (b) Different people can have different notions of what is illuminating.

So illumination by itself doesn't make a very good organisational tool for mathematics. In the end, doing mathematics is not just about convincing onself that things are true; the point is to advance the knowledge of the world around us, not just the knowledge inside our own head.

The circle of truth

I am going to describe mathematical activity in terms of moving around between these three kinds of truth:

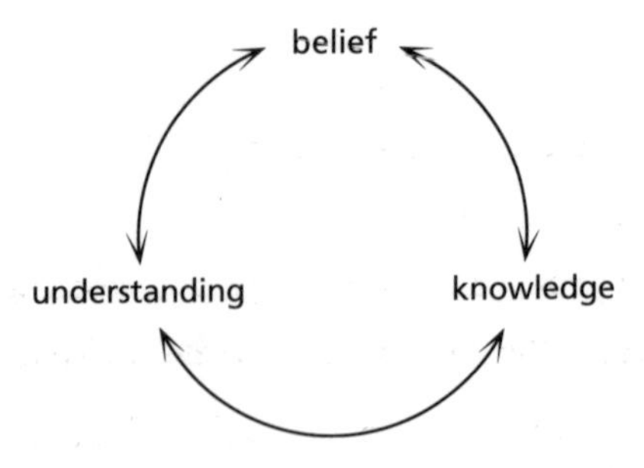

In maths, knowledge comes from proof – we know something is true by proving it. Usually we think that the big aim of doing maths is to prove theorems, that is, move things into the 'proved' area. But I think the deeper aim is to get things into the 'believed' area – believed by as many mathematicians as possible. But how do we do that? If I have proved something is true, how do I really come to believe it? It's as if I have some sort of illuminating reason for believing it, rather than just following the proof through step by step. However, once I believe it, how do I convince someone else? I show them my proof.

We need the proof to enable us to move from the realm of *my* believed things to anyone else's:

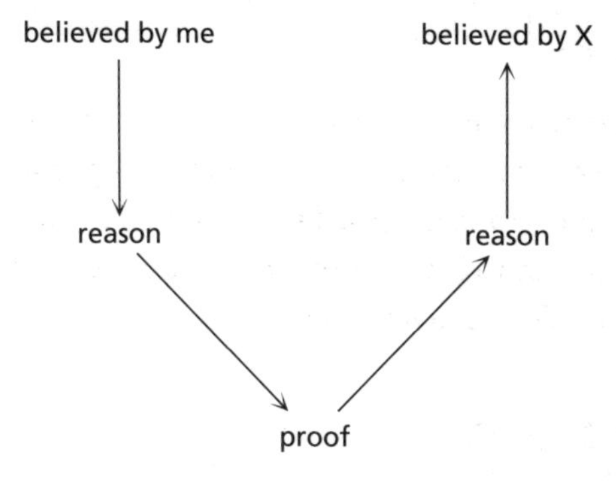

So the procedure is:

* I start with a truth I believe that I wish to communicate to X.
* I find a reason for it to be true.
* I turn that reason into a rigorous proof.
* I send the proof to X.
* X reads the proof and turns it into a convincing reason.
* X then accepts the truth into his realm of believed truth.

In fact, it's not so much a circle of truth as a valley; attempting to fly directly from belief to belief is inadvisable.

We've all seen people try to transmit beliefs directly, by yelling. So if transmitting beliefs directly is unfeasible, why don't I just send the reason directly to X, thus eliminating what are probably the two hardest parts of this process: turning a reason into a proof, and turning a proof into a reason?

The answer is that a 'reason' is harder to communicate than a proof.

I think that the key characteristic about proof is not its infallibility, but its sturdiness in transit. Proof is the best medium for communicating my argument to X in a way that will not be in danger of ambiguity, misunderstanding or distortion. Proof is the bridge for getting from one person to another, but some translation is needed on both sides.

When I read someone else's maths, I always hope that the author will have included a reason and not just a proof. When this does happen, the benefits are very great. Unfortunately a lot of maths is taught without any attempt at illumination. Even worse, it's sometimes taught without any explanation at all. But even if it is explained, not every explanation is illuminating. For example, we mentioned earlier that when you learnt how to solve something like

$$x + 2 = 5$$

you might have been told, as I was: 'You take the 2 over to the other side of the equals sign and the plus becomes a minus.' This gives

$$x = 5 - 2$$

so

$$x = 3.$$

This is correct, but unilluminating. Why does that trick with the equals sign work? Apparently one way of teaching this is that the plus sign moves through the equals sign and the vertical bar gets stuck, so $+$ turns into $-$. This is a pretty absurd way of teaching it, because then what happens when you send a minus sign through? A very unilluminating explanation.

At least in the UK and US, many people grow up feeling great antipathy towards maths, probably because of how they were taught it at school, as a set of facts you're supposed to believe and a set of rules you have to follow.

You're not supposed to ask why, and when you're wrong you're wrong, end of story. The important stage in between the belief and the rules has been omitted: the illuminating reasons. An illuminated approach is much less baffling, much less autocratic, and much less frightening.

But is there always an illuminating explanation for every piece of maths? Probably not, just as there is not an illuminating explanation for everything that happens in life. Some things so incredible or tragic happen that no explanation is possible.

Category theory seeks to illuminate maths. In fact, category theory could be thought of as the universal way of illuminating maths – it seeks to illuminate, and that's all it does. That's its role. That's the glass slipper into which it perfectly fits. I'm not claiming category theory explains everything in mathematics, any more than mathematics explains everything in the world.

Mathematics can seem like an autocratic state with strict unbending rules that seem arbitrary to the citizens of this 'state': the pupils and students. Schoolchildren try to follow the rules, but are sometimes abruptly told that they have broken a rule. They didn't do it deliberately – most students who get some maths questions wrong didn't do it on purpose, they really thought they had the right answer. And yet they're told they've broken the law and will be punished – being marked wrong feels like a punishment to them. Perhaps it is never really explained to them what they did wrong, or perhaps it was not explained to them in an illuminating way that could actually make sense to them. As a result they don't know when they will next be found to have broken a rule, and they will creep around in fear. Eventually they'll simply want to escape to a more 'democratic' place, a subject in which many different views are valid.

* * *

'Knowledge is power', or so the adage goes. But understanding is more powerful power. We have moved on from the age when knowledge was a secret, passed around in mysterious books that could only be deciphered by a small number of people. We have moved on from the age when there were so few books that even those who did know how to read them were at the mercy of those who owned them, the age when students seeking knowledge had to gather around somebody who would read the book out loud to them, a 'lecturer' – remember that the word 'lecture' comes from the act of reading, not the act of pontificating to an audience. Anyway, we have moved on from that age.

We are now in the age where information is everywhere. Literacy rates still leave room for improvement, but most adults can read, and in some countries most of them have access to the internet. Many of us essentially have the internet in our pocket at all times. Knowledge is no longer a secret. But understanding is still kept a secret, at least in mathematics. Students of all levels are shown the rules but kept in the dark about the reasons. We encourage children to ask the question why, but only up to a point, because beyond that point we might not understand it ourselves. So we stifle their quest for illumination to match our own inability to provide it. Instead of being afraid of that darkness, we should bring everyone to the edge of it and say: 'Look! Here's an area that needs illumination.' Bring fire, torches, candles – anything you can think of that will cast light. Then we can lay down our foundations and build our great buildings, cure diseases, invent fabulous new machines and do whatever else we think the human race should be doing. But first of all we need some light.

ACKNOWLEDGEMENTS

I am deeply grateful to so many people that I'm beginning to wonder if it wouldn't be better to thank nobody at all rather than omit people, but perhaps that's taking logic to extremes in a way that I don't advocate.

So first I'll thank my friends and collaborators in the category theory research community. My conversations with them, mathematical or otherwise, are a continued source of inspiration and excitement. Some of them will recognise where they are thanked implicitly in the text. I would also like to thank my non-mathematician friends who have been sufficiently curious about my work to have given me years of practice at describing it by way of analogy, anecdote and anything other than technicalities.

Credit is due to my friends and family whose curious incidents or insightful observations on life I mentioned in the text, with or without their names: my mother, father (who also took the bagel, slinky and knot photos), sister, little nephews Jack and Liam, Tyen-Nin Tay, Noo Saro-Wiwa, Brandon Fogel, James Martin, Mike Mitchell, Celia Cobb, Karla Rohde, Marina Cronin, the late Professor Philip Grierson, James Fraser, Jim Dolan, Amaia Gabantxo and first year student Frank Luan, who wrote me the letter from which I quoted at the start.

I would like to thank all my students who ever failed to understand something so that I had to explain it better, my agent Diane Banks, Nick Sheerin and Andrew Franklin at Profile, and TJ Kelleher and Lara Heimert at Basic Books. I also thank Sarah Gabriel, for always being a beacon when my brain was getting foggy, and I thank Jason Grunebaum, Oliver Camacho and James Allen Smith for their love.

Chapter 5 is dedicated to Gregory Peebles, who is so moved by the concept of a torus.

Finally I'd like to thank Kevin × 2, Natalie, Slava, Ryan and Tim at the Travelle in Chicago, for keeping me nourished while I did the final edits, all small children for brightening the world, and everyone else.

There, I think that covers it.

INDEX